Autonomic Nervous System & Homeopathy

Personalized Nutrition Concept Depicted in Homeopathy & Ayurveda

Dr Govindsing Girase

ISBN 978-93-5610-403-7
© Dr Govindsing Girase 2022
Published in India 2022 by Pencil

A brand of
One Point Six Technologies Pvt. Ltd.
123, Building J2, Shram Seva Premises,
Wadala Truck Terminal, Wadala (E)
Mumbai 400037, Maharashtra, INDIA
E connect@thepencilapp.com
W www.thepencilapp.com

DISCLAIMER: *The opinions expressed in this book are those of the authors and do not purport to reflect the views of the Publisher.*

Author biography

Dr. Govind Girase is a Homeopathic Specialist with a mission. He aims to eliminate the fear of disease suffering from the minds of people by affecting a complete cure in as many cases as possible. He endeavors to make his services and treatment available at affordable costs.

Dr. Govind practices in Kalyan, specializing in Asthma. Dr. Govind holds a MSc. Pharmaceutical Medicine postgraduate degree, Bachelor of Homeopathic Medicine and Surgery. This is a 5 1/2-year course in the medical field associated with homeopathic medicines and diploma in Clinical Research (PDCR). Dr. Govinds's homeopathic research papers has been published in International Journal of Allied Medical Sciences and Clinical Research (IJAMSCR) and also written a book 'Autonomic nervous system and homeopathy'.

A holistic view of health Although current medical practice recognises that there are connections between mental, emotional and physical conditions, the dominant approach is to break the whole down into parts with separate diagnostic labels. Different drugs are then prescribed for each seemingly unrelated symptom. Most medicines have side effects which add to the patient?s troubles - a search on the Internet will reveal pages of them. To counteract these side effects a new set of drugs may be given. Thus suffering is not ultimately decreased at all but actually increased. Homeopaths use specially prepared, potentised medicines from which toxic effects have been removed. Homeopaths look at the big picture: all the patient?s symptoms are grouped together and understood to represent the whole disease. When making an individual assessment homeopaths look at ancestral health, family issues, conception and gestation, birth and childhood, whilst at all times noting the ways in which the patient manifests uniqueness. Job and family life, hobbies and holidays, relationships, responses to life?s challenges and everything that denotes individuality are used to add to the picture of health and its breakdown into disease. One of the reasons that homeopathy is the second largest system of medicine in the world today, aside from bringing health benefits without side effects or the dangers inherent in major interventions, is that it offers this holistic approach to health and healing. Holism implies that the parts in a system relate interactively; that body, mind and spirit are all aspects of a whole being, and that all individuals are involved in a web of interrelationships with the world, its mineral, plant and animal constituents. Apropos of which, the healing agents used in homeopathy derive from mineral, plant and animal sources and are used according to the principle of like cures like.

Homeopath

Education

BHMS - KDMGS Gudhe Shirpur Dhule - 2010

Master of Science in Pharmaceutical Medicine - M.H.U.S (Maharashtra University Of Health Sciences - 2013

PGDCR - Haffkine Institute Parel - 2011

CONTENTS

1. Personalized Nutrition

It is observed and commonly understood that; the changes in the life styles and food patterns have created more complex health issues. The term personalized nutrition has been established in traditional medicine. A person's character, immunity, and metabolism had been experimented and practiced with natural principles. Action of food on health and disease state needs to be understood from individualization concept. The food plays a vital role in the development of fetus up to the old age Hippocrate quoted "Let food be thy medicine and medicine be thy food" – The future of Holistic Healthcare. Any living organism on earth manifest vital force - energy balance for maintenance of harmony. Susceptibility varies in different individuals, Susceptibility can be increased, decreased or destroyed 'One man's meat is another man's poison' The food metabolism is essential for maintenance of internal environment.

Food habits have been changed with industrialization however farming and fruit and vegetable season depends on environmental changes. There is changing need of bodily energy with respect to season; whereas processed food is available all the times with fixed amount of energy of macro and micronutrients. These modified lifestyle increases physical and mental stress different life style related disease are on the rise which require critical evaluation of personality and his interaction to confront surrounding condition.

Modern nutritional science has confirmed the benefits of traditional Homeopathic and Ayurvedic herbs. The foundation of personalized nutrition is laid by our ancestors of traditional medicine. Potent phytochemicals and mineral elements protect the body from carcinogens, depression, allergens, and metabolic changes and stimulate immune cells.

Protection from carcinogen:

There are many research studies suggesting food has anti-cancer effect in disappointingly ineffective cancer patients 1. Sulphur is a commonly prescribed remedy in Homoeopathy; Food containing Sulphur has potential effect in cancer concomitant therapy. Allium family onion and garlic, contains variety of sulphur compounds 3. Organic and inorganic sulphur compound studies stated that it may be useful as prevention and inhibition of cancer cell proliferation2.

Allergy:

In early childhood food related allergic symptoms are more common and it is a sign of developing immunity 4. Recent approach of treating food allergy with probiotics has shown consistent improvement especially selective strain specific bacteria according to type of allergy or allergens 4.

Immunity:

Defense provided by functioning of immunity can be modulated by correct food. Homoeopathy considers intellectual behavioral signs and symptoms in drug proving, as a person's character is reflected as the immunity from environment Successful application of food as immunomodulation based dietary recommendation has long being employed successfully in traditional medicine.

Depression:

Magnesium helps to improve symptoms of depression 6. Magnesium, Vitamin B-12 and Folate has been studied in Randomized Control Trial (RCT) with positive results Increase intake of high calorie food is associated with deregulation of stress hormone cortisol, and circulatory corticosteroid association with metabolic changes and obesity has been elucidated 7.

Autonomic Nervous System (ANS) serves as communication between external environment and homeostasis. Metabolic and Endocrine activities, Sexual behavior, micturition, respiration, intestinal movement, cardiovascular system are governed by ANS Emotional arousals like Increase or decrease in heart rate, gastric motility, cutaneous blood flow, blushing, piloerection, and sweating are governed by Sympathetic Nervous System (SNS). It prepares body to utilize metabolic resources; parasympathetic nervous system promotes building up of metabolic reserve, and every person is different according to his reaction to surrounding.

Sympathetic Nervous System governs development of asthma complaints, Sympathetic activity increases in bronchial asthma to balance parasympathetic over activity 8.

Finding from muscle sympathetic nerve activity (MSNA) studies, suggest that inverse relationship between MSNA and Cardiac output 9. Sympathetic and Parasympathetic Nervous System infiltrate tumors to contribute early stage of gastric and prostate cancer 10. Glucose deregulation is associated with autonomic dysfunction 11. SNS has role in pro-inflammatory process, and impairment in immunity is cause of infection 12.

Could there be a way to design personalized diet plan which improves a person's immunity, 'Personalized Medicine' — a medicine specifically designed keeping in mind a patient's biology. Scientist's are working on cutting edge technology that brings us closer to modern medicine. This is because predominant modes of drug development rely on testing on animals and a very small group of human beings who very likely belonged to different ethnicities, genders, age groups.

The composition of nearly many medicinal plants is known today Body is composed of food, drink and air, the constituents are used to form body tissue, enzymes, hormones blood ect.

If nutritional goal is not achieved it causes long term deficiency which leads to increased disease susceptibility. In case of chronic and lifestyle diseases Ayurveda Yoga Naturopathy, Unani and Homeopathy AYUSH system needs support from nutritional aspect. Randomized Controlled Trial has proved role of nutrition in following health condition.

Folic Acid - Birth Defect

Calcium - Osteoporosis

Omega Fatty Acid - Atherosclerosis

Vitamin D - Calcium Deficiency

Vitamin C - Cold and Flue

Iron - Anemia

Allium Family Plants - Cancer Progression

Aspargus Recemosus - Lactation

Probiotics - Gut Health

Curcumin - Osteoporosis

Fats:

Several RCT has demonstrated anti-inflammatory properties of essential fatty acid. There are two basic type of fats Saturated and Unsaturated If an individual lack enzymes for fat metabolism, It might damage tissues of brain, spleen, liver, and heart.

Proteins:

Proteins are essential for physical development Bioavaibility of protein differs in plant and animal source. In case of deregulation of protein metabolism harmful toxic substance buildup in body, sever inflammation, muscle contraction, and insulin secretion is affected.

Carbohydrates:

Carbohydrates are most nourishing energy efficient fuel source These are consumed by body tissue in form of glucose and fructose. Constant supply of glucose is required for metabolic function of brain, Red Blood Cells (RBC) and inner portion of adrenal medulla. Disturbed sugar metabolism is needs to be studied to understand insulin resistance.

Minerals:

Minerals are important for enzymatic function Deficiency of minerals depends on kind of daily work; these micronutrients are required as person's life style Long term deficiency results in adverse changes in normal physiological functions.

Vitamins:

Vitamins are essential for metabolism, nerve impulse, anemia, fatigue, and heart disease Constitution wise investigation and supplementation is helpful to re-balance body vital energy Vitamin B is important for cognitive function; especially Vitamin B-12 plays paramount role in Deoxyribonucleic Acid (DNA) production is well studied. Frequent consumption of water soluble vitamins are must as they eliminated in urine and sweat. Fat soluble vitamins are stored in fatty tissue.

Action of Homeopathic Medicine on autonomic nervous system is confirmed by Dr Akalpita S Paranjpe using Medical Analyzer System at Electronics Division, BARC, Mumbai, India 13 Similarly association between physiological patterns and Tridosha theory, Prakriti parikshan and Metabolism in Ayurveda is well explained with knowledge of complex physiological reactions 14, 15 .

Food supplements are not only essential for physical equilibrium, but also helps maintain physiological reactions. Vital force can function appropriately only if body system is properly supplied with micro and macro nutrients.

In course of time Dr Hahnemann and our Homoeopathic ancestors has left a rich wealth of well experimented and clinically verified curative medicine. A deliberate evaluation of particular sign and symptoms in a good number of drugs has been elucidated. Dr Hahnemann proved drug Sepia on basis of Similimum A patient who had developed chronic ailments because of absorbed Indian ink in stomach; he had habit of wetting his painting brush with saliva It will be fruitful Endeavour to expand our knowledge about action of Homoeopathic medicine and new therapeutic field of Lifestyle and Incurable diseases. In terms of Homoeopathy chronic ailments or disease is considered as dysfunction of Harmony in energy, i.e. vital force This dysfunction of vital energy presents clinical signs and symptoms. Totality of sign and symptoms gives Individualized picture of proved drug In the conditions like asthma, arthritis, infertility, and mental Illness where allopathic research is struggling to minimize systemic side-effects, Homoeopathy assures Ideal cure.

Different homeopathic remedies indicate exhibit different physical constitution; In general metabolism differs in every person with type of food Carbohydrate, Protein, Fat 16. Carbohydrates are essential for energy level balance, protein is important for muscle mass and fats are efficient form of energy, brain is fattiest organ 60% of fat. Lean and thin sulphur, tall and slender phosphorus, obesity in thuja are common observation in clinical practice. The different constitutional remedies states psychological makeup which indicative to distinguished metabolism profile. In ideal case elegant carbohydrate metabolism is useful in protein utilization, and capable to develop utilization of fat stores. If sugar metabolism is unbalanced, eating more calories from fat and protein source will cause excess of unutilized energy.

In case of Sulphur remedy especially suited for lean person with stooped shoulder, in which complex carbohydrate, protein, and fat metabolism is deranged Symptoms of gastro intestinal track includes FAINT FEELING IN ABDOMEAN AT 11 AM; WEAK AMPTY ALL GONE Painful and difficult defecation, constipation alternates with diarrhea 17.

Kent Repertory > Generalities Sec > Page Number 1362

Bread agg

(Complex Carbohydrate): Ant-c , bar-c , Bry , carb-an , caust , chin , clem , coff , crot-h , crot-t , kali-c , merc , nat-m , nit-ac , nux-v , olnd , ph-ac , phos , Puls , ran-s , rhus-t , ruta , sars , sec , sep , staph , sul-ac , sulph , teucr , zinc , zing

Meat (Animal Source Protein):

Carb-an , caust , colch , cupr , ferr , kali-bi , lyss , mag-c , mag-m , merc , ptel , puls , ruta , sil , staph , sulph , ter

Fat agg:

Acon , ant-c , ant-t , ars , asaf , bell , bry , carb-an , carb-s , Carb-v , caust , chin , colch , Cycl , dros , eupho , ferr-ar , ferr-m , Ferr , hell , hep , ip , kali-ar , kali-c , kali-chl , kali-n , mag-c , mag-m , meny , merc-c , merc , nat-a , nat-c , nat-m , nat-p , nit-ac , nux-v , phos , ptel , Puls , rob , ruta , sep , sil , spong , staph , sulph , Tarax , thuj , verat

Food and supplements containing Sulphur compound Amino Acid – Methionine, Cysteine, Taurine, Vitamins – Thiamine, Biotin, Methylsulfonylmethane – wheat germ, onions, brussels sprouts, garlic, asparagus, kale, legumes, and Glucosamine Sulphate, Melatonin, cruciferous vegetables.

Modern nutrition science is at preliminary phase of individualization, we need to apply knowledge from traditional text authenticated by modern clinical studies. Role of food in immunity development is not studied as per traditional knowledge, if studied preventing fetal infections and genome related disease would be achievable goal Addressing health with gut microorganism is the beginning – Probiotics Common cause of disease is weak defense system; moreover good immunity and intellect are required for survival on earth. Homoeopathy states individualization clearly, which regulates bodily response to environment stimulus or stress.

Scientists are exploring complex physiological function and role of food to improve longevity. The dynamic organization of structural, biochemical function and enormous co-ordination of organ function offers scope to understand Ayurveda conceptualize theory. The conventional western medicine views body from structural perspective whereas Ayurveda mentions human body system and different health approaches with different terminologies and metaphor.

Homeopathic medicines are prescribed on basis of totality (Constitution, Temperament, Thermals, Miasm, and Diathesis) need to be studied in relation to Sympathetic Nervous System. Ever since Dr Samuel Hahnemann proved drug Cinchona deliberate evaluation of physical sign and symptoms, psychological profile of homeopathic medicines has been elucidated by proving on basis of Similia Similibus Curantur.

Doctrine of drug proving is assessment of safety (Monitoring of serious adverse event) and efficacy (Action of remedy physical and mental Symptoms) by proving it in low doses in healthy human being. There are distinctive concept of Ayurveda such as Dravyaguna, Bajikarna, Rasayan Tantra, Kayachikitsa Tantra, Ayurveda states 'Preservation to Health of Healthy Person and Treating Ailments with Breaking Causative Factor of Pathogenesis' Ayurveda is not restricted to few Samhitas, but scattered through other mythological text like Smruti, Smriti and Puranas. The much overlooked Homeopathic system of medicine has immense benefits of immunity development. Homeopathic constitution wise food habits and aversion are experimented in details. Its principal states Ideal cure, varying degree of disease susceptibility and protection from natural diseases.

Many healing plants in Ayurveda and Homoeopathy are to be studied as individualized essential food to prevent diseases. Plants do not have immunity, they develop metabolites that are toxic to herbivorous animals, fungi, bacteria and viruses.

Table 1: Following are few Ayurveda/ Homoeopathy medicinal plants mentioned in Food Safety Standard Authority of India Gazette Notification 6 January 2017, its primary use as food supplement.

Botanical Name	Benefits
AllinmSativum	Anti-Inflammatory, Anti-Bacterial, Rheumatism, Skin Affection, Alopecia
Ginkgo Biloba	Memory Enhancer and Anti-Oxidant
ZingiberOfficinalis	Bronchitis, Hyperglycemia, Diarrhea
Panax Ginseng	Debility, Lumbago, Sexual Excitement
Glycyrrhiza Glabra L	Anti-inflammatory, Expectorant, Anti Allergic
Hypericum Perforatum	Neuralgia, Scar, Ulceration Wound
Curcuma Longa	Antiseptic, anti-Arthritis, Anticancer, Antiseptic
Allium Cepa	Hypoglycemia, Ant atherosclerosis
Aloe Barbadensis	Dilates Blood Vessels, Wound Healing, Constipation
Cassia Angustifolia	Purgative
Centella Asiatica	Spasmolytic, Anti-Anxiety, Liver Affection, Gall Stone, Dropsy

Ayurveda Theories

Tridosha Theory

Sapta Dhatu Theory

Panchbhuta Theory

Fundamental Principals in Homoeopathy

Law of Similar

Simple Remedy

Minimum Dose

Potentization

Vital Force:

Drug Proving

Theory of Chronic Disease

Enormous research has been conducted to evaluate plant structural properties and active ingredients. There is further scope to evaluate concept of personalized nutrition, with Homeopathic natural law, theories, and principals. Homeopathic constitution wise food advice will synergize bodily resistance. Ayurveda conceptual tacit knowledge about Panchbhuta, Sapta Dhatu, Tridosha, Ttva can be implemented in understanding of broadly experimented Homeopathic medicine and action of food on body.

From drug discovery point of view research in genomics is restricted due to Cost, chemical-biological safety and toxicity, ethical issues. Genome has a complex biochemical and physiological function, though human genome mapping has been completed, different response in different patient to same drug is not completely understood.

Allopathic drug response differs to different individuals with same disease, complex biochemical and physiological function of genes, gene therapy holds promise for cure in human disorder and lifestyle diseases. Homoeopathy has been warmly recognized and highly accepted by Indian society Traditional Ayurvedic vaidya prescribed particular diet according to seasons with prakriti examination.

Concept of personalized nutrition will be particularly very helpful for Homeopathic practitioner to treat patients with

Family history of lifestyle diseases

Patients with risk of developing diseases

New born and adolescent were immunity is in developing stage

Avoid recurrence of symptoms or short relief in case of chronic diseases

Reference

1. Skyler B. Johnson, MD et. al. Complementary Medicine, Refusal of Conventional Cancer Therapy, and Survival Among Patients With Curable Cancers, JAMA Oncol. 2018;4(10):1375-1381. doi:10.1001/jamaoncol.2018.2487

2. Lee J et al Anti-Cancer Activity Of Highly Purified Sulfur In Immortalized And Malignant Human Oral Keratinocytes Toxicol In Vitro 2008 Feb;22(1):87-95 Epub 2007 Sep 1

3. Mark Durenkamp Luit J De Kok, Impact Of Pedospheric And Atmospheric Sulphur Nutrition On Sulphur Metabolism Of Allium Cepa L , A Species With A Potential Sink Capacity For Secondary Sulphur Compounds Journal Of Experimental Botany, Volume 55, Issue 404, 1 August 2004, Pages 1821–1830

4. Food-Related Symptoms And Food Allergy In Swedish Children From Early Life To Adolescence Jennifer L P Protudjer, Mirja Vetander, Inger Kull, Gunilla Hedlin, Marianne Van Hage, Magnus Wickman, Anna Bergström Published: November 15, 2016 Https://Doi Org/10 1371/Journal Pone 0166347

5. Kuitunen M, Probiotics And Prebiotics In Preventing Food Allergy And Eczema Curr Opin Allergy Clin Immunol 2013 Jun;13(3):280-6 Doi: 10 1097/ACI

6. Rajizadeh A, Mozaffari-Khosravi H, Yassini-Ardakani M, Dehghani A, Effect Of Magnesium

Supplementation On Depression Status In Depressed Patients With Magnesium Deficiency: A Randomized, Double-Blind, Placebo-Controlled Trial Nutrition 2017 Mar;35:56-60 Doi: 10 1016/J Nut 2016 10 014 Epub 2016 Nov 9

7. Nieuwenhuizen AG1, Rutters F Physiol Behav The Hypothalamic-Pituitary-Adrenal-Axis In The Regulation Of Energy Balance Physiology And Behavior 2008 May 23;94(2):169-77 Doi: 10 1016/J Physbeh 2007 12 011 Epub 2007 Dec 23

8. Th Pricila Devi, W Kanan, Th Shantikumar Singh, W Asoka, Benjamin L; 2017, A Study Of The Sympathetic Nervous System In Bronchial, Asthma, Journal Of Medical Society / Sep-Dec 2012 / Vol 26 | Issue 3

9. Michael J Joyner, M D *, Nisha Charkoudian, Ph D , And B Gunnar Wallin, M D , Ph D ; 2011 July, The Sympathetic Nervous System And Blood Pressure In Humans: Individualized Patterns Of Regulation And Their Implications, National Institute Of Health Hypertension Author Manuscript; PMC 2011 July 1

10. Claire Magnon, 2015, Role Of The Autonomic Nervous System In Tumorigenesis And Metastasis, Molecular & Cellular Oncology 2:2, E975643; April/May/June 2015; Published With License By Taylor & Francis

11. MERCEDES R CARNETHON PHD; DAVID R JACOBS JR PHD, STEPHEN SIDNEY MD

MPH, KIANG LIU PHD; 2003, Influence Of Autonomic Nervous System Dysfunction On The Development Of Type 2 Diabetes, DIABETES CARE, VOLUME 26, NUMBER 11, NOVEMBER 2003

12. Georg Pongratz* And Rainer H Straub; 2014, The Sympathetic Nervous Response In Inflammation, Pongratz And Straub Arthritis Research & Therapy 2014, 16:504 Http://Arthritis-Research Com/Content/16/6/504

13. Mishra N1, Muraleedharan KC, Paranjpe AS, Munta DK, Singh H, Nayak C , An exploratory study on scientific investigations in homoeopathy using medical analyzer J Altern Complement Med 2011 Aug;17(8):705-10 doi: 10 1089/acm 2010 0334

14. Dey S, Pahwa P, Prakriti and its associations with metabolism, chronic diseases, and genotypes: Possibilities of new born screening and a lifetime of personalized prevention J Ayurveda Integr Med 2014 Jan;5(1):15-24 doi: 10 4103/0975-9476 128848

15. Frederick T Travis Robert Keith Wallace Dosha brain-types: A neural model of individual differences, J Ayurveda Integr Med 2015 Oct-Dec; 6(4): 280–285 doi: 10 4103/0975-9476 172385

16. James T Kent, Clara L Kent, Kent Repertory of Homeopathic Materia Medica, Enriched Indian

Edition reprinted from Sixth American Edition Generalities Section, Food – Page Number 1362, 2004

17. S K Subey, Text book of Materia Medica including alles keynote for easy explanation, Publisher Books and Allies Pvt Ltd 1999

2. Philosophy

Ayurveda has basically classified human personalities into 3 different categories Vata, Pitta and Kapha. None above mentioned has one single dosha predominant. It is the blend of all three doshas but collectively one dosha dominated over other. The science of Ayurveda has been practiced from ancient time in India. Ayurveda is now very popular worldwide and it has been doing well as alternative method of disease prevention and treatment worldwide. If you do not understand ancient scriptures then it's very difficult to understand its importance. The main concepts in Ayurveda are sapta dhatu, tridosha, sthula and sukshma. The skepticism about science is very dangerous due to non-comprehension of these concepts. The detail scriptures give us a broad complementary concept to physiology and disease pathophysiology. Theses scriptures give an idea about human body interaction with environment. In Ayurveda tridosha theory vata pitta and kapha is very clearly mentioned. It is not merely a disease phase. A group of molecule that suggest over all body composition and functioning. The tridosha theory is based on physiological function energy production, digestion and excretion. The concept of sthul and sukshma (macro and micro) is a subtle legacy in Ayurveda. It gives logical reasoning in subtle anecdote. The body is group of 5 elements earth, water, fire, air and sky. These 5 elements are like redundant substances which continuously form new substance (organ development) to release energy. The process of forming new substance and release of old used one with release of energy; Food substance is catalyzed by tridosha particles vata pitta and kapha. Tridosha operated through five elements, energy utilization, and excretion of unused particle and release of energy.

Energy Utilization - Vata – Psora – Absorption – Earth Water

Nourishment - Pitta - Psychosis – Digestion - Fire

Segregation - Kapha – Syphilis - Excretion – Air Sky

Even if we cannot see tridosha through operating we can conceptualize from daily living activities and physiological reasoning. Earth is mainly energized by sun, moon and atmospheric air. There are numerous living and non-living substances present on earth which are influenced by these three energies. Earth is solid element, water is liquid, we cannot touch fire or air but we can see fire, we can feel presence of air. Sky means hollow space. Water and moisture holds solid particles together. Heat or fire is required to strengthen water solid mixture (Mix food and water in stomach). Air and Space gives body exact shape. The hollowness and dimension presents amount of air present in it (stomach). There are many living organism on earth however human have capability to conceptualize. Any flaws in understanding these basic concepts will further restrain to understand disease pathophysiology and tridosha presence all over body.

Kapha – Air Sky –Atmospheric Air – Syphilis

When considering boys as five elements Kapha contains sky and air elements. Kapaha is oily and sticky substance which is very slow in action. Kapha personalities are very slow in thinking; they are usually absorbed into their own thought process. Liver store micro and macronutrients and segregate these macro and micronutrients in this process cough get excreted into lungs. Kapha represent autonomic nervous system of upper part of body

Sphere of action:

Chest, Head, Joints, Gums, GI track, Sweat glands

absorbed, salivation, laziness, chills, asthma, cough, sleepiness, deglutination, intestinal movement, excretion of stool, urine, semen and fetus.

Calm and composed

Complacent nature

Firm judgment

Slow paced

Slow digestion

Sleep deeply

Large body frame

Slow movement

If he gets angry it remains for very long time

Blends easily with opposite quality personalities

Kapha is required to increase efficiency of lungs for continuous breathing. It is suggested to eat less fat in asthma cases. In throat it works to facilitate deglutination, oily substance present in throat helps swallow food particles; without sticking to inner wall. It is not just an oily substance but force that helps movement of food bolus from mouth into GI track. Kapha in head gives calmness to five senses (smell, taste, vision, touch and hearing) It helps keep bran calm. In joints kapha gives lubrication which helps in movement of joints. Force that expels excreta is main function of kapha. White coated tongue is a very good example to denote kapha prakriti. It helps in food swallowing, kapha provide protection from cold weather due to extra amount of adipose tissue. The function of kapha particles can be understood from its location and its function.

Pitta – Fire –Sun – Sycosis

Pitta helps in body nourishment helps segregate micro and macro nutrients from stomach and intestine and its storage in different compartment such as protein into muscles, sugar into liver, rbc and fats into adipose tissue. In Ayurveda pitta is compared to sun which helps mix water, digestive juice bile into food particle in process which facilitate essential nutrients absorption.

Pitta sphere of action:

Umbilicus, Stomach, temperature, skin eyes, acidity, hunger, thirst

Anguish

Cannot tolerate heat (Psoriasis)

Cheerful

Social

Quick digestion

Intolerant at work

Get along with people very well

Cannot tolerate hunger

Digestive enzymes

Brilliance

Pupil constriction

Vission

Vata – Earth Element – Moon- Psora

Heart and lung covering – Energy utilization – Organ development (creation)

Vat means energy converted from digestive particles. The power of intestinal absorption and digestion improves nourishment to body. These stored nutrients are then converted into energy which is released at cellular level; energy required to perform daily task for survival on earth. Psora represents lower autonomic nervous system.

Sphere of action

Large Intestine, rectum, hair, bones, movement, constipation, generalized weakness, flatulence, tremor, forgetfulness, loquacious, mental disorder, obsessive compulsive disorder, panic attack, irregular periods and arthritis.

Active – cannot seat at one place

Can not concentrate

Agitated

Less sleep

Eat poor

No good digestive function

Malnourished

Cannot tolerate heat

Lack of decision power

Never happy with other person

Imaginary world of movies, comics and games

Homeopathic drug proving states constitution, immunity, and metabolism is well experimented science with natural principal. Time tested major constitutional medicine protects from allergy, depression, carcinogen, and metabolic changes, immunity development. Concept of Ideal cure is rooted by Master Dr. Samuel Hahnemann, it's an art overlooked by modern medicine scientist. They are now relying on personalized medicine which has failed to integrate psychological aspect of deranged health. India is homeopathic hub of the world. Potential theories of environmental interaction and psychological response theories are being accepted with better results. It is very difficult to get uniform result where interdisciplinary role of physiological reaction, anatomical variation, non-specific immunity, and metabolic response plays a vital role. Modern medicine proving with strong targeted response is a totally different science from the classical homeopathic drug proving.

Every living creature has to protect himself from the foreign pathogens and has to maintain internal environment. Industrialization has affected environment, most comprehensive environmental interaction is through food. A passive example is psychological reaction to a trigger. Our conscience is the key for judging right from wrong. Progressive biological response to every species is different to unique stimuli, biodiversity is essential for survival. example - Symbiosis, parasitic infection.

Sensation method is based on a person's environmental interaction in plant and animal and mineral kingdom. Since, then we improved precision in prescribing medicine, there is further scope of research in finding relationship between remedies. Now a day's remedy relationship is widely utilized for treating the disease conditions; however complete restoration of vital force can not be elucidated with complete understanding of deranged energy. Case taking - Physical, psychological signs and symptoms observed with respect to miasm, seasonal changes. The safe and effective method of observing vital force is ANS medical analyzer.

Thus it is important to incorporate remedy relationship effectively. An important remedy relationship can be stated with respect to metabolism and Hering's law of cure.

"Healing occurs from within out, from the head down, and in the reverse order in which the symptoms have appeared."

Deranged vital force affecting fat metabolism need attention at the beginning followed by remedies complementary protein and carbohydrate metabolism.

Remedy

Metabolism

Constitutional Makeup

Sulphur

Sugar -Evolving

Source of Information

Phosphorus

Protein – Deranged

Natures Enormous Diversity Information

Silicea

Protein – Evolving

Haughty in Knowledge Implementation, Fascisms

Alumina

Fat – Deranged

Professionalism, Multitasking

Mercurius

Fat - Evolving

Social Values

Food plays important role in enzymatic activity, hormonal balance, and various physiological reactions. Fat and proteins are more complex form of energy source than sugar. If sugar metabolism is deranged then individual will struggle to survive evolution process.

Does it suggest importance of carbohydrate diet?

Why lycopodium tree has survived process of evolution?

Why lycopodium does not follows sulphur?

Why renal stones are common finding in Lycopodium and Sulphur constitution?

Carbohydrate diet is essential for existence in today's era of information. It gives a sense of reality in unrealistic world of expectation and insecurity. Proteins are essential for perfection in implementing idea, where strenuous physical activity is required with difficulty in mental labor, mistakes in reading, writing cannot bear to perform cognitive task for long period. Fat metabolism is associated with hurried and worried for idea protection, nurturing it and outcome oriented, and frequent repetition of experiments in life to for social goal, much related to cognitive ability of a person.

An Idea to improve metabolism with exercise is general, however fat, sugar, and protein metabolism specific exercises are partially known. Sugar is constant source of energy. Proteins are linked to stamina and strenuous physical activity. Essential fatty acids are morbid source for cognitive function.

Frequent change of remedy is dangerous for patient; more simplified approach is to prescribe step by step carefully observing symptoms that relate to type of deranged metabolism in sequence of fat, protein, and carbohydrate metabolism. One essential factor linked with metabolism is seasonal changes and type of infection.

Winter-Fat metabolism-Alumina- Thuja (Inflammation Limbs feel as if made of wood)-eye inflammation-viral infection

Rainy -Protein metabolism-Silica-Nux vomica (frequent desire for stool)-Bacterial infection

Summer -Sugar metabolism-Sulphur- Cina (Hungry after full meal) –Cough and Dry cough with sneezing-parasitic infection.

As long as diseases have power to make people sick, we continue to investigate derangement of vital force and causative factor. Such investigations should be focused on environmental interaction. Element indicates fight for survival – Sympathetic Nervous System, biochemical salt reassure the sources available (rest and digest) Parasympathetic Nervous System, plant remedy psychological symptoms has both sources for growth and fight for protection. Animal enjoys sources, act as a predator max benefits from a situation.

We generally overlook ornamental information a person is collecting to survive on earth and focus on subjective responses of past experience. Can we arrange homeopathic remedies with respect to site of action on autonomic nervous system? It will evaluate harshly criticized remedy relationship. Undoutably many researchers work hard to establish relationship between mineral, plant and animal kingdom. There is indeed plenty of evidence to assure it.

Lachesis-Lycopodium-Ferrum Met-Phosphorus

An assumption that disease can be cured with any potency is needed to be tested as we need to achieve cure with perfect harmony of vital force.

Factors affecting choice of dose:

Disease complexity

Disease severity

Duration of disease development

Major constitutional remedies

Site of ANS

Disease complexity:

A person with deranged sugar metabolism is more susceptible for natural diseases. Essential response of immunity development can be achieved with small doses. Approximate quality of response to complex protein and fat metabolism can be improved with high doses.

Disease severity:

To trace response of remedy low dose is essential, however a physician is assured about totality of symptoms where higher potencies will give finer result. ANS affecting minor changes in physiological reaction require low doses example sneezing, bloating. High potencies are more simplified form of crude drug with precise action. Example: Constipation ~ lycopodium 30C, Renal Stone ~ lycopodium 1M.

Duration of disease development

Duration of disease development correlated with its pathophysiological aspect or degree of metabolism disorder. More convenient high dose is required for long term adaptation.

Major constitutional remedies

As compared to plant and animal kingdom similar threshold therapeutic result can be obtained with low dose of mineral kingdom. High dose of animal and plant kingdom gives favorable result.

Example: Animal Kingdom $\geq$ Plant Kingdom $\geq$ mineral Kingdom

Type of ANS

Symptoms related to sympathetic nervous system responds to low dose and parasympathetic nervous system need higher dose.

The concept of Psora Syphylis and Psychosis is closely aligned with Sugar, Protein and Fat metabolism.

Psora-Sugar Metabolism

Psychosis- Protein Metabolism

Syphilis- Fat Metabolism

Immunity and Food:

Channelizing immunity in right direction with food is a progressive natural phenomenon in every individual. Immunity development is succession of various challenges (threat) posed by environment and bodily response to it is new path of investigation. The term disease is in state of chaos for modern medicine physicians.

On basis of observation immunity is evolving state of disease metabolism a double edge sword. One good example is evolving state of protein metabolism and tuberculosis. Can we improve metabolism with sensitization to specific food or specific microorganism? Dynamic food in health restoration. Principle cause of increase disease susceptibility is indigestible and inappropriate quality of food. There are ever increasing number of invention in modern medicine, however scientific treatment rationale for genetic disorder like thalassemia is at the initial stage of development. It will help to minimize incidence of genetic disorder, rheumatoid arthritis, psychological disorder, cancer and lifestyle diseases. Genetic disorders drug research has to go through a hazardous process of evaluating safety and efficacy. Homeopathic constitution specific food genuinely benefits patients. It states pathophysiology disease with role of specific biomolecule. Varied response of medicine acting on specific organs system documented in detail. Drug proving has stated medicine interaction with physiological process. The individualization concept is essential for conduct of modern medicine clinical trial, safety and efficacy evaluation in specific group to obtain uniform result. Autonomic nervous system establishes meaningful relationship of disease pathophysiology. Radical improvement in disease prevention is an achievable goal. Diabetes and hypertension require urgent intervention with homeopathic constitution specific diet. Over the past few year there is increasing controversy in genetic cloning as well as safety of nano medicine. it will enable regulatory bodies in drug research with individualization concept.

Psora:

Psora is derived from greek word itch, most common chronic mism. Psora is considered as mother of all diseases, it is particularly related to ectoderm. Its functional disorder consists of intolerable itch. On suppression of local itch symptoms disappear, however state of derangement progresses inward this is the state of latent psora. Patient may look apparently healthy with latent state of increase disease susceptibility.

The secondary manifestation of psora is more dormant form, susceptible to external stimulus. Anger is causes deviation from healthy state of psora. Ectoderm is related to itch and protection, associated with skin diseases.

Sycosis:

The symptoms are first produced by suppressed emotions, followed by inflammatory process. A state of transition subdued quarrelsome nature into state of jealousy, deceitful behavior.

Syphilis:

The miasm states functional and emotional state of distruction. Reserved energy broken by intellectual task or huge amount of emotional turmoil to disintegrate vital force. Syphilis is susceptible to action of mercurius and natrum mur.

Syphilis	Subordinate Food Supplement
Suspicious	Polyunsaturated Fatty Acid
Irritable	Nitric Oxide
Jealous	
Excessive proliferation	**Probiotic:** Bifido Bacterium Strain, Lacto bacillus Strain
Slow recovery	
Leucorrhea	
Dryness	**Other:** Essential Fatty Acid, Walnut, Sea Food, Meat, Sodium Salt, Cocoa, Dark, Chocolate, Mushrooms, Ginger, Apple Cider Vinegar, Cinnamon, Pine Apple, Oranges, Honey, Grape Seed Extract, Crocus Sativus, Astazanthine, Zin Sulphate
Acrid discharge	
Deformed nails	
Cold weather agg.	
Exaggerated cell or tissue growth	(Susceptible to: Parasite= Bacteria = **Virus**)
	(Derranged Metabolism: (Sugar= Protein =**Fat**)

Psora Symptoms	Subordinate Food Supplement
Alert	**Avoid:** Complex Carb, Cookies, Fermented Food, Cake
Quick	
Physically active	
Mental fatigue	**Probiotic:** Enterococcus Strain
Fits of anger	
Emotional disturbance	**Other:**
Fear of death	Alpha Lipoic Acid, Corn, Fruits, Rice, B complex vitamin, Chondroitin Sulphate
Complaints relived by heat	
Hand Feet Burn	DMSA, Garlic, Glucosamine Sulphate
Itching Dry Skin	Magnesium Sulphate, Methionine
Constipation	
Hungry After Eating	Milk Thistle, MSM, N Acetyle Cystein
Intelligent	Taurine, Nitric Oxide, Artichokes, Cruciferous Veg
Dirty	
	(Susceptible to: **Parasite** = Bacteria = Virus)
	(Metabolism: **Sugar**)

Syphilis	Subordinate Food Supplement
Suspicious	Polyunsaturated Fatty Acid
Irritable	Nitric Oxide
Jealous	
Excessive proliferation	**Probiotic:** Bifido Bacterium Strain, Lacto bacillus Strain
Slow recovery	
Leucorrhea	
Dryness	**Other:** Essential Fatty Acid, Walnut, Sea Food, Meat, Sodium Salt, Cocoa, Dark, Chocolate, Mushrooms, Ginger, Apple Cider Vinegar, Cinnamon, Pine Apple, Oranges, Honey, Grape Seed Extract, Crocus Sativus, Astazanthine, Zin Sulphate
Acrid discharge	
Deformed nails	
Cold weather agg.	
Exaggerated cell or tissue growth	(Susceptible to: Parasite= Bacteria = **Virus**)
	(Derranged Metabolism: (Sugar= Protein =**Fat**)

3. Genetic Disorder

Gene Therapy holds promise for cure in human disorder life style and infectious diseases like HIV. From Homeopathic point of view psychological with respect to flight and fright regulation gives us new avenues to explore future studies.In chronic disease like obesity, Diabetes, Hypertension, IHD, Arthritis, network of genes fails to perform, however Homeopathic medicine prescribed on basis of totality (Constitution, Temperament, Thermals, Miasm, Diathesis) need to be studied in relation to Sympathetic Nervous System. Synthetic Hormone response differs to different individual, at least one episode of hyperglycemia in 86% of cases. Who receive high dose of steroid. Steroid induces 60 to 80% of insulin resistance depending on dose and type of steroid. Synthetic hormone has ability to depress anterior pituitary. Synthetic Hormone and Sympathomimetics has systemic side effects, it affects growth and development of body.

Thalassemia is an inherited blood disorder, in which body makes abnormal type Hb protein. Hb protein has two major type alpha globin and beta globin, defect in gene leads to the abnormal formation of one of this protein. In case of thalassemia minor there is faulty gene from one of the parent. Thalassemia patients have no symptoms with small RBC size.

Incidence:

Thalassemia is blood disorder majorly affecting Southeast Asia, south Asia and Mediterranean region. Thalassemia affects 56,000 infants born annually with major thalassemia1.

Thalassemia cases are predominantly observed in south china. In china about 20000 cases of Thalassemia are born annually. It is common hereditary blood disorder in china 47.48 million thalassemia carriers 2.

% of HbF at Different Stages 3

Normal newborn: 71 ± 7.7%

Premature Infant: 81 ± 4.7%

Post Mature Infant: Approx. 55%

Normal Adult 2%

Erythropoiesis: Total 16 RBC maturation process from one elytroid stem cell. These cells are produced after pronormoblast or committed stem cell undergoes cell division. 10% erythrocytes die before they mature. In case of megaloblastic anemia affects process of erythropoiesis and fails to deliver 50% of cells 4. The stage of RBC development includes mitotic division with nucleus development and subsequent changes in cytoplasm.

Why cells fail to reach circulation?

In spite of erythropoietin activity why RBC do not mature?

Is there any factor that affects cell division?

Following diagram shows RBC 6 stage development

Stage 1 Pronormoblast 12 to 14 micro gram

Stage 2 Basophilic normoblast / Proubricyte stage

Stage 3 Polychromic Erythroblast / Polychromatographic Normoblast

Stage 4 Orthochromic normoblast

Stage 5 Reticulocyte

Stage 6 Erythrocyte

Alpha Thalassemia:

These individuals may develop shortness of breath and fatigue. In case of thalassemia major Bart hydrops fetalis is the sever type, Its less sever type is known as HbH.

Bart hydrops fetalis sign and symptoms includes severe anemia, enlarged spleen, enlarged liver, genital abnormalities, increased body fluids, and defect in heart development.

HbH has sign and symptoms in milder form: yellow eyes, jaundice, hepatomegaly, and splenomegaly.

Gene deletion is most common in alpha thalassemia

Alpha thalassemia 1: Both alpha chain gene deletion (Common in South East Asia)

Alpha Thalassemia 2: One alpha gene deletion (Common in South East Asia)

HbH Disease: moderate anemia Alpha 1/ Alpha 2 three gene deleted

Barts Hb: most severe form may lead to intrauterine death, 4 genes are deleted Alpha1/ Alpha1 type.

Genotype Alpha1/ Alpha Osmotic fragility is seen with mild anemia. 2 genes deleted.

Genotype Alpha2/ Alpha: 1 gene deleted with very mild clinical condition.

Genotype Alpha2/ Alpha 2Vitamin B 12 deficiency in absence of macrocytic anemia.

Beta Thalassemia:

Enlargement of liver and spleen is due to excessive RBC destruction. Intense bone marrow hyperplasia causes bone expansion. Treatment includes Regular blood transfusion 2 to3 units every 4 to 6 weeks to maintain Hb 11 mg/dl, , Regular folic acid supplementation with 5 mg daily dose. Iron chelation therapy used to prevent iron overload. Vitamin c 200mg daily for excretion desferrioxamine iron overload. Spleenectomy may be needed, Bone marrow transplant from HLA matches siblings. Endocrine treatment if required to stimulate delayed puberty, as replacement.

Diagnosis is based on following parameters, sever hypo chromic, microcytic anemia with raised reticulocyte percentage, basophilic stippling in blood film. Hb electrophoresis vessels absence or almost complete absence of HbA with almost all Hb circulating HbF. Alpha/beta chain synthesis studies on circulating reticulocytes shows marked increase in alpha/beta ratio that is absence or reduced synthesis of beta chains.

It is caused by reduce or absence synthesis of beta globulin chain. It causes skeletal abnormalities, poor growth and hemolytic anemia. Patient with beta thalassemia develops iron overload by age of 30 which leads to cardiac complication. A study published in medical principles and practice journal 2014, conducted at Mumune hospital turkey investigated metabolism in beta thalassemia minor. Out of 194 patients 92 enrolled in Beta thalassemia group and 102 enrolled in control group without beta thalassemia 5. The fasting insulin, fasting glucose, HDL and triglycerides were higher in study group. The high glucose and insulin level suggest deranged glucose metabolism.

Other studies have also correlated increased glucose and insulin levels. Bahar et al reported that development of insulin resistance in subjects with beta thalassemia was more frequent due to oxidative stress and hemolysis 6. Tong et al observed normal glucose tolerance with insulin resistance and high fasting glucose level in thalassemia minor subjects 7. Increase in

HDL and Triglyceride is associated with low LDL 8, 9.

This is most frequent mRNA defect, reduced Beta globulin chain. There are three different patterns.

mRNA not detectable

Nonfunctional B globulin mRNA

Structurally abnormal mRNA globulin

Lepore Thalassemia:

High concentration of HbF, during meiosis there is unequal crossing between alpha and beta globulin.

F Thalassemia or Delta Thalassemia:

This disorder is more commonly seen in Greece. Defective alpha and delta chain production

Thalassemia Intermedia Some cases of Beta thalassemia are of moderate severity Mild clinical condition which does not need transfusion. The patient with Thalassemia Intermedia presents a moderate deformity enlarged liver and spleen, extra medullary erythropoiesis and symptoms of iron overload in adulthood. Hypochromic microcytic blood picture MCV, MCH,and MCHC all are very low with mild or no anemia Hb 11 to 15 mg/dl. Raised Hb A2 >35% confirms diagnosis 10.

Treatment:

Blood transfusion on regular basis.

Folate supplement

Iron Supplement

Bone marrow transplant.

Patient on regular blood transfusion may require Chelation therapy to decrease iron load.

Genetic therapy for thalassemia patients is under phase 3 clinical trial, focused on haemopoetic stem cell transplant. It's difficult to get HLA compatible donors.

There is little recent reliable approach in Ayurveda herbs without complete recovery. There are anecdotal reports on Hydroxyurea and wheatgrass to enhance Thalassemia Intermedia cases outcome.

Homeopathic clinical studies have shown encouraging results 5. It is claimed that homeopathy offers supportive treatment to thalassemia homeopathy address to root cause of diseases, which decrease frequency of blood transfusion required. Immunity sensitization with homeopathic medicine is required in suspected cases 11.

The cases of thalassemia need homeopathic treatment to enhance glucose metabolism. There is incapacity to metabolize carb, protein, and fat. These characteristics are transferred from parents. The affected sugar metabolism leads to defect in bone development specifically cell division. Sulphur and related remedies can be used to enhance metabolism.

Genetic information is more specifically transferred in Autonomic nervous system functions. There is urgent need to conduct study to evaluate this hypothesis. Large RCT has proved action of ANS on different organ sites. Parasympathetic 10th Cranial(Vagus) nerve serves as communication between immune system and brain. VIOME a us based firm analyses gut bacteria, fungi, virus, to find out human body composition (Individualization). Gut microbe colony specific diet recommendation.

Autonomic nervous system serves as mediator to balance harmony between external environment and homeostasis. ANS governs major metabolic function, cardiovascular function, sweating, and urination, utilization of energy. Parasympathetic nervous system helps building metabolic reserve.

Reference

1. Darlison M et al Global epidemiology of hemoglobin disorder and derived service indicator Bulletin of world health organization. 86(6) 480-487.

2. Li CGet al Thalassemia incidence and treatment in China with special reference to Shenzen cityand Guangdong province Hemoglobin 33(5) 296-303

3. Clinical Hematology R D Estahan Sixth edition John Write and Sons Limited 1984 Published by English language Book Society (Haffkine Institute library book number 8097 page number 23).

4. Principles of hematology Peter J Haens, Layola Marymount university, Wm C. Brown publishers, (Haffkine Institute library, Book number 9876 Chapter 4, Year 1995).

5. Sinan Kirim et.al. Is beta thalassemia minor is associated with metabolic disorder, Medical principle and practice karger publishers Sep 2014, 23(5): 421-425.

1. Adele Bahar et. al. relationship between beta-globulin gene carrier state and insulin resistance, Diabetes Metabolic Disorder Journal 2012; 11:22 PMCID PMC 3598161.

1. Peter Tong, C reactive protein and insulin resistance in subjects with Thalassemia minor and a family history of diabetes, diabetes care journal 2002 Aug; 1480:1481.

2. Mario Maioli et. al. plasma lipid in beta thalassemia minor, Atherosclerosis 1989 Feb; 75(2-3):245-8.

3. M Hashemieh et. al. Bangladesh Med. Res. Counc Bull 2011 Apr; 37(1):24-7. Hoferbrand, P.A.H. Moss, J. E. Pettiit second edition, Essential of Hematology.Wiley, 18-Oct-2001.

4. Antara Banerjee, Sudeepa Basu et. al. Can Homeopathic medicine bring additional benefits to thalassemia patients on hydroxyurea therapy?Encouraging result of homeopathic treatment. Evidence based complementary alternative medicine, march 2010; 7(1): 129-136.

5. L Anguilar Lopez et. al. Thalidomide therapy in patients with thalassemia major. Maxico. Blood Cell Mol. Dis.2008 Jul-Aug.41(1):136-7.Epub 2008 Apr 24 DOI10: 1016/j.bcmd.2008.03.0011.

1. Additional Reference

2. Anand S Dutta, John P Griffin; 2007, Discovery of New Medicne, The Text Book of Pharmaceutical Medicine Fifth Edition Page 4

3.   Clore Jn, Turby Hay L, 2009; Glucocorticoid induced hyperglycaemia, Endocrine Practice 2009, Jul- Aug 15 (5); 469-74.

4.   Strohmayer EA, Krakoff CR. 2011; Glucocorticoid and Cardiovascular Risk factor Endocrine Metabolism, Endocrinol Metab Clin North America. 2011 Jun;40(2):409-17, ix. doi: 10.1016/j.ecl.2011.01.011.

5.   Fong AC, Cheung NW; 2013, The high incidence of steroid-induced hyperglycaemia in hospital, Diabetes Res Clin Pract. 2013 Mar;99(3):277-80. doi: 10.1016/j.diabres.2012.12.023. Epub 2013 Jan 5.

6.   Th Pricila Devi, W Kanan, Th Shantikumar Singh, W Asoka, Benjamin L; 2017, A study of the sympathetic nervous system in bronchial, Asthma, Journal of Medical Society / Sep-Dec 2012 / Vol 26 | Issue 3

7.   Michael J. Joyner, M.D.*, Nisha Charkoudian, Ph.D.*,†, and B. Gunnar Wallin, M.D., Ph.D.; 2011 July, The sympathetic nervous system and blood pressure in humans: individualized patterns of regulation and their implications, National institute of Health Hypertension. Author manuscript; PMC 2011 July 1

8.   Claire Magnon, 2015, Role of the autonomic nervous system in tumorigenesis and metastasis, Molecular & Cellular Oncology 2:2, e975643;

April/May/June 2015; Published with license by Taylor & Francis

9. MERCEDES R. CARNETHON PHD; DAVID R. JACOBS JR. PHD, STEPHEN SIDNEY MD MPH, KIANG LIU PHD; 2003, Influence of Autonomic Nervous System Dysfunction on the Development of Type 2 Diabetes, DIABETES CARE, VOLUME 26, NUMBER 11, NOVEMBER 2003.

10. Georg Pongratz* and Rainer H Straub; 2014, The sympathetic nervous response in Inflammation, Pongratz and Straub Arthritis Research & Therapy 2014, 16:504 http://arthritis-research.com/content/16/6/

4. Remedy Relationship

Seven Chakras

Muldhara: Survival instinct, inner safety, Internal freedom, Selfishness, seek transformation

Swadhishthana: Imagination, Visual Hallucination, Seduced by temptation, try to controle other people life, verbalizing Joy, Sexual Organ, Desire for pleasure, Anger, Lust

Manipurka: Personal identity in society, arrogance, ineffectual, Greed, self-confidence, memory intelligence, Self-awareness, ego

Anahat: Alert and attentive, love, kindness, Family oriented cloth, shelter and food, Creativity, Love, Affection, Grief, Hear voices,

Vishudhi: Accepts light and dark side of life, communication, insecurity, happiness, seeking truth.

Ajna: Command center of body, logical reasoning, boundry between human and divine, merged into universe.

Sahasrara: Ego death, intelligence.

Embryology

Genetic information coded in sperm and ovum represents autonomic nervous system activity in respective sex. Embryology development is biological process this encoded genetic information is used in differentiating organ with respect to autonomic nervous system activity. Human development can be seen in different aspect; morphological, psychological, physiological, metabolic changes. Both sexes suggest difference in autonomic nervous system activity, male represents sympathetic nervous system and female as dominant para sympathetic nervous system. More sympathetic activity in mother gives birth to male child. Child sex differentiation is not just based on observation it is based on conceptual theory from natural phenomenon and environmental changes (full moon – sympathetic activity) the environmental changes are governed by planetary movement, these changes affects autonomic nervous system. The transfer of characteristics from parents to child is taken as transfer of panchmahabhuta, tridosha and saptadhatu trait.

Cell differentiation and organogenesis are nothing but parent's interaction with environmental interaction by autonomic nervous system. If there is prolonged grief or anger in parent before fertilization the characters get transferred to child. There are five main feelings lascivious, anger, grief, attraction, fear. Every feeling presents respective panchamahabhuta earth, water, fire, air, sky. These feelings provide protection from environment. Feelings represent parasympathetic nervous system (resources e.g. veratrum – money, pulsatilla-emotional bonding, nux- idea implementation, Lycopodium –

Intelligence, Ignatia- self-protective behavior) Sympathetic nervous system denoted work, concentration, will power (Sulphur –ego, phosphorus-indifference, silicia-work anxiety, alumina-perfection, merc-social benefits). Sympathetic nervous system is inner aspect of a personality the intellectual aspect of a person. Parasympathetic nervous system is effect of environmental, social, family life on body (Mag – absence of mind, ferrum – haughty, kali-obstinate, calcarea-protective and natrum-passion self-control). The ultimate aim of a person should be to do best for him/her using all resources. Our view of genetic science is needed to be broadened. In general parasympathetic nervous system is active in females and male sympathetic nervous system is active. There is reserved parasympathetic in males and sympathetic energy in females. Emphasis given to ailments from is very general we need to classify remedies as per metabolism (psora-sugar, syphilis-fat, and sychosis-protein). It's quite easy to understand vital force in embryological layers it moves from ectoderm to endoderm. Morphological and psychological development takes place from outer layer to inner endoderm. Tridosha and panchmahabhuta theory enable us to understand organ and psychological development with respect to season. Embryological development with respect to mental capability is recent field of research. In our materialistic world psychological profile is need to be understood from physical characteristics.

Oocyte is well protected in uterus with sources of nourishment. Sperm cell functions to survive by penetrating ovum. Sperm and ovum develops in opposite direction; ovum specifically develops into cell cytoplasm, and sperm has quality of cell nucleus. The basic phenomenon in fertilization is polarization. The polar differentiation in fertilized ovum represents opposite forces. The process of gamete fusion takes place at fertilization stage, the chromosomes become diploid. In this process sexual identity is characterized. In some species like kangaroo, deer environmental changes have definitive effect on fertilization process.

Embryonic life follows biological clock. The organ differentiation is governed by environmental condition through autonomic nervous system. Mothers nutrient intake influence mental capabilities and psychological profile of child. The gastro intestinal track anatomy and physiology and animal instinct differs in every species. The dynamic aspect of embryological layer development opens possibility to understand miasm. Human body has high level of intelligence however in animal's senses and organs are developed to function a specific task like flying, griping, smelling, running.

Feelings like fear, anger, grief mental symptoms protect us from injury to vital force. Metabolic and protective phenomenon in frog, snake, cuttlefish, insects and bird differs. The phase of metabolic profile represents specific autonomic nervous system activity. Emotions are like animal instinct that acts like protective ectoderm earth frog lascivious, water snake anger, fire cuttlefish grief, air insect attraction, Sky bird fear. In Ayurveda human body is considered as last animal in evolution process that is eligible for salvation.

If we consider our body as cell, digestive track is nucleus and emotional makeup and other organs are in cytoplasm. Gastrointestinal track represents sky element. This sky element is devised into of two imaginary half first at inward force and outward force that is sympathetic and para sympathetic. Inward force is required for self-development and out word force is infinite opposite to Inward force.

The process of games fusion takes place at fertilization stage at this stage chromosomes become deploying after fusion. The first completely developed organ is from mesoderm tissue I.e. Heart. Mesoderm tissue also forms vascular system, cartilage, tendons, bones, muscles. Gastrulation is primary to development of central nervous system. Central nervous system communicates with environment at concepts level where as gastrointestinal track takes food from environment to maintain homeostasis. Excreted products are organic in nature to maintain natures life cycle. Thus gator intestinal track has dynamic relationship with environment where as somatic nervous system or brain is responsible for protection. Thus ingestion and perception directly correlates with fright and digest. Excretion and protection correlates with fight and

re

Central Nervous System
Brain
Parasympathetic
Division
Sympathetic
Division
Constricts pupil
Dilates pupil
Stimulates tear glands
No effect on tear glands
Strong stimulation of salivary flow
Weak stimulation of salivary flow
Inhibits heart, dilates arterioles
Accelerates heart, constricts arterioles
Constricts bronchi
Dilates bronchi
Stimulates stomach motility and secretion, stimulates pancreas
Inhibits stomach motility and secretion, inhibits pancreas and adrenals
Stimulates intestinal motility
Inhibits intestinal motility
Contracts bladder
Relaxes bladder
Spinal cord
Stimulates erection
Stimulates ejaculation

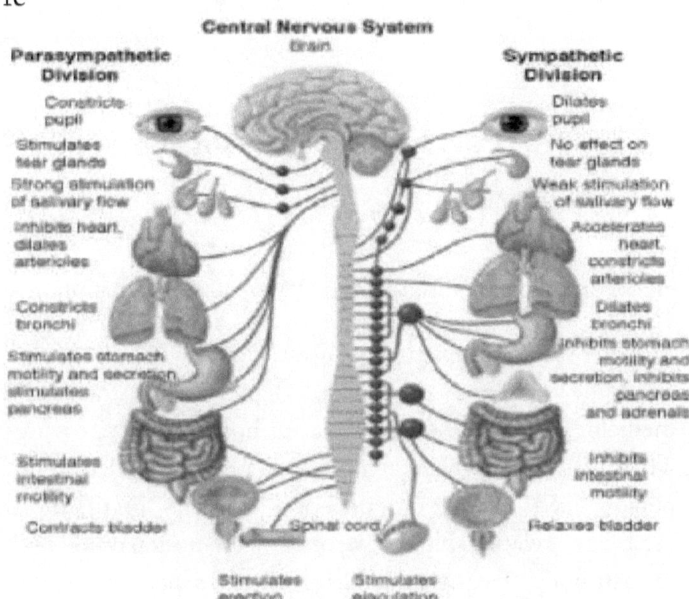

5. Psychological Disorder

Homeopathic Constitution and Personalized Nutrition

Modern medicine research is useful source of information regarding explicit knowledge depicted in traditional medicine. We have huge number of homeopathic provings which is comprehensible with vast knowledge of personalized nutrition concept. We are gratified to establish relationship environmental interaction with human body. Treatment is based on key aspects of patient Immunity, metabolism and cognition. Homeopathic medicine is consistently showing benefits in anxiety and depression related disorder. In Allopathic system generalized anxiety disorder is treated with following class of medicine.

SSRI

Venlaflaxine

Nefazodone

Matrazapine

Moderately effective

TCA

MAOI

Essential fatty acid potential benefits has been studied previously which is related to deranged fat metabolism. The generalized anxiety is more commonly seen in children and adolescents. Antidepressants are preferred clan of drug with good efficacy. The selective neurotransmitter re-uptake inhibitor prevents re absorption of specific bio-molecule into the nerve cells of brain however there is no surety that how it regulates depression.

Anxiety Neurosis Treatment Available in Allopathy

There are psychological therapies, cognitive behavioral therapy and antidepressant occasionally with antipsychotics has proven benefits. Benzodiazepine may help alleviate insomnia and anxiety but not depression with dependency and withdrawal issue. The SSRI are first line of treatment for most anxiety and depressive disorder with dose titration in case of side effects.

SSRI acts by increasing levels of serotonin at nerve junction which improves nerve function. Commonly available SSRI antidepressant in market

Ventafaxin

Bupropion

Nefazodone

Mirtazapine

SNRI inhibit absorption of Serotonin and norepinephrene. Which include duloxetin, venlafexin, desvenlafaxin, levomilnacipran. Desvenlafaxin ER.

NDRI which inhibits reuptake of norepinephrine and dopamine. It includes only one drug that is bupropion.

Tetracyclic works by preventing binding norepinephrine and serotonin specific undesired receptors in adjacent nerve cells thus improving quantity at synapse. Example Asamoxapine, maprotiline, mirtazapine.

SARI: Serotonin antagonist reuptake inhibitor prevents undesired serotonin receptor binding in adjacent nerve. Example nefaxodone and trazodone.

Psychological disorders are very are need tobe studies with respect to specific emotions like anger and anxiety, Fear and Agoraphobia.

Ziziphus Jujuba

It is selected after critical evaluation with natural principles and knowledge depicted in ayurveda. It is Ideal food supplement for anxiety; stress may or may not be associated with depression. It restores health and can be used as complementary therapy to allopathic and homeopathic medication. There are no know side effects or drug interaction, safe food ingredient from food safety standard authority of government of India approved gazette. It is best suited with allopathic drugs as action is more precise location similar to Bupropion1 as mechanism of action of prescribed medicine is explored with modern scientific studies. The précised action of medicine is predicted to improve nor epinephrine and dopamine function. A homeopath carefully prepares case report form to evaluate mental, physical and environmental response, draw a totality of symptoms which is compared with the drug picture for remedy. Ziziphus Jujuba is safe and effective for the treatment of anxiety associated with depression safe and effective in children as well as adults 3. It can be more effective in women. More effective in remitted anxiety and depressive disorder may or may not be associated with disability pain1. It augments effect of selective serotonin reuptake inhibitor. There are no side effects of sexual dysfunction or weight gain. It's efficacious in treating post-traumatic stress disorder.

Treatment Available in Traditional Medicine for Anxiety Disorder

Though treatment for anxiety is available in Ayurveda and homeopathy, it is not very frequently used due to limited clinical evidence and non-uniformity in outcome of treatment. 5000 years of science offers herbal remedies for alleviation of anxiety and depression. Modern clinical studies have confirmed the calming effect of herbal Ayurveda remedies.

Homeopathic Medicine for Anxiety and Depression:

Sepia

Natrum Mur

Ignatia

Arsenicum Album

Phosphorus

Ayurveda Medicine for Anxiety and Depression:

Kawa

Chamomile

Valerian Root

Arctic Root

Ziziphus Jujuba prescribed in powder form 15 gm daily for 2 months along with homeopathic med. It augments effect of selective serotonin reuptake inhibitor. There are no side effects of sexual dysfunction or weight gain. Its efficacious in treating post-traumatic stress disorder. Dopamine gets activated by nor adrenaline and adrenaline. Secretion of hormone is specific to region of autonomic nervous system. Sulphur Androgens, Phosphorus – Adrenaline, Silicia – Pancrease, Alumina – Thyroid, Merc – Pituitary. Ziziphus Jujuba complements action of Phosphorus related autonomic nervous system. Ziziphus Jujuba with vitamin D, Carbohydrates, resveratrol, walnut (omega fatty acid), homeopathic Ferrum salt ect. Synergistically improve hormonal balance and metabolism.

Reference: -

1. Timothy R. Berigan, the Many Uses of Bupropion and Bupropion Sustained Release (SR) in Adults Prim Care Companion J Clin Psychiatry. 2002; 4(1): 30–32.PMCID: PMC314381 PMID: 15014734.

2. Eric W. de Heer et al, The Association of Depression and Anxiety with Pain: A Study from NESDA, PLoS One. 2014; 9(10): e106907. Published online 2014 Oct 15. doi: 10.1371/journal.pone.0106907PMCID: PMC4198088 PMID: 25330004

3. Sarris J. Herbal medicine for treatment of psychiatric disorder : systemic review Phytother Res 2007 Aug. 21(8); 703-16.

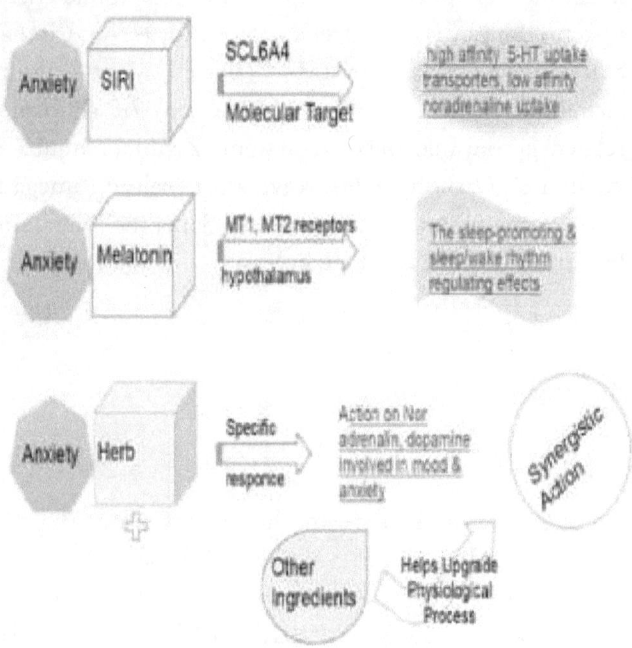

6. Nutrition

Homeopathic Healing and Nutrition

Current level of Global Nutraceutical Market is $ 294.49.6 billion 2017, growing at CAGR compound rate 8.8%. It is expected to cross $578.23 billion by 2025. In US nutraceutical market functional food segment is growing fastest. Japan, US, Germany, Italy, France, UK and Italy are key markets. Indian Nutraceutical Market is growing with annual increase 21% according to study on nutraceutical segment by Associated Chambers of Commerce and Industry of India 2018. The study statistics states that India's current nutraceutical market is $ 4 billion 2017 and it is expected to reach $18.38 billion by 2025 1.

We can develop a disease specific formulation based on disease patho-physiology. Modern clinical studies have confirmed varying degree of disease susceptibility and role of deranged function bio molecule responsible for it. We can validated disease pathway and conceptualize knowledge depicted in alternative medicine. Our aim should be to enhance the physiological function in disease state to synergistic ideal homeostasis condition based on authenticity and precision.

Clinical studies

Dr. Akalpita Paranjpe former BARC scientist has confirmed the action of plant and mineral kingdom on ANS. Variable dose of 200 to 1M homeopathic potency was first time with valid scientific experiment. heart rate and blood flow was studied with Sulphur, Phosphorus, Gelsemium, Pulsatilla and Aconite 2.

Association between physiological patterns and tridosha theory, prakriti parikshan and Metabolism in ayurveda is well explained with knowledge of complex physiological reactions 3.

Team of IIT Bombay Scientist has proved that extreme Homeopathic dilution of gold, copper and iron retains nano particles 4. Homeopathic medicine retains the property of crude substance characterized by nuclear magnetic resonance, this study states that homeopathic dilution can not be considered as placebo and there is an evidence gelsemium, copper memory in dilution which can be recorded. Neuronal circuit of autonomic nervous system are responsible for regulation of physiological state in dynamically changing post postpartum environment 6. Immune response in target cells or lymphoid tissue is a result of diverse signaling pathway which include neurotransmitter, neuro modulator, cholinergic anti inflammatory and adrenergic nerves. Autonomic nervous system regulatory link between immunity associated changes with respect to natural diseases 7. A study published in Indian Journal of Research in Homeopathy states effect of homeopathic medicine in behaviour disorder in autistic children, There was significant improvement in neuo psychological symptoms such as hyperactivity reduction, communication difficulty, sensory impairment, behavioral dysfunction 8. Modern nutrition science is at preliminary phase of individualization, we need to apply knowledge from traditional text authenticated by modern clinical studies. Role of food in immunity development is not studied as per traditional knowledge, if studied preventing fetal infections and genome related disease would be achievable goal. Addressing health with gut microorganism is the beginning. Common cause of disease is weak defense system; moreover good immunity and intellect are required for survival on earth. Homeopathy states individualization clearly, which regulates bodily response to environment stimulus or stress.Scientists are exploring complex physiological function and role of food to improve longevity.

Epilepsy

Epilepsy is one of the most common paroxysmal and heterogeneous neurological disorder affecting over 42 million people worldwide with distinct symptoms, etiology and prognosis. According to estimate worldwide 8 per 1000 people have epilepsy 1.

The treatment outcome for epilepsy with anti-epileptic drug may vary, even between patient and seemingly the same epilepsy syndrome. It is estimated that 20 to 25% of patients with epilepsy fail to achieve good control with anti-epileptic drugs 2. The cause and manifestation with epilepsy are so diverse and multifaceted that it impinges on virtually every aspect of modern biological science, from molecular biology to vocational rehabilitation. It could be because of brain damage caused by a difficult birth, a severe blow to head, a stroke, or an infection to brain such as meningitis.Various studies in last 15 years have shown genetic alteration associated with epilepsy syndromes. There are mainly due to the mutation in the genes regulating sodium channel protein. These defective ion channels remains open for a long time thus causing the neuron to get hyper excited.

The epileptic seizures can be of various types including partial seizure, generalized seizure or status epilepticus. Partial seizure or focal seizures arise from a particular area of brain and affect body at specific location. Generalized seizure on the other hand arises due to generalized abnormal electrical activity in the brain and in turn involves various subtypes like clonic, tonic, generalized clonic-tonic, absence seizures and so on status epilepticus is the most life threatening form. A case of status epilepticus would be a generalized tonic-clonic seizure patient suffer from hypoxia or anorexia.

The effective management of epilepsy is very important for the patient to maintain good health and it can be achieved by diet and allopathic drugs. To choose the most suitable AED for individual patient it is necessary to have detail knowledge about the properties of seizure. . Phenytoin, Phenobarbitone, Carbamazepine, and Valproic Acid are some of the first choice anti-epileptic drugs which have proved successful in management of seizure.

Phenytoin is hydantoin compound related to the barbiturates that are used for treatment of seizure. It is an effective anticonvulsant for the chronic treatment of clonic-tonic or partial seizure and acute treatment of generalized status epilepticus. The anti-seizure activity of phenytoin is related to its ability to inhibit repetitive firing of action of potential caused by prolonged depolarization of neuron. Additionally phenytoin stop the spread of abnormal discharge from epileptic foci there by decreasing the spread of seizure activity throughout the brain. At the cellular level, the mechanism of action of phenytoin appears related to the ability to prolon the inactivation of voltage activated sodium ions channels and reduction of ability of neurons to fire higher frequencies. Phenytoin side effect at higher doses includes depression, drowsiness, fatigue, confusion, lethargy, nystagmus. Personalized medicine is defined as medical care for each patients unique conditionand had its root in understanding of disease pathophysiology. The concept of personalized medicine root back to 1500 BC, Pharmacogenetics is study of inner individual variation in DNA sequence related to pharmacokinetics and pharmacodynamics 3, 4.

Gene regulating sodium ion protein denotes parasympathetic nervous system related to sodium salts and fat metabolism.

Epilepsy is associated with aggression observed in preictal, ictal and postictal phase 5. Interictal diasphoric disorder: Agression is a deffencive behavior indicating self-protection, interictal diasphoric disorder describes aggressive and irritable behavior inbetween two episode of seizure 6. Phosphorus and related group of remedies (lachesis, pulsatile, ferrum, stramonium, hyoscamus) indicates anger, aggression and anxiety. Anger is the potential trigger factor in epilepsy. An uncontrollable aggression is often found associated in epilepsy patients. Frequents attack of anger displaces a person into anxiety, depression. Study published in acta Neurologica Scandinavica aims to assess effect of anti-epileptic drug carbamazepine analogue. Patients with history of psychological disorder were excluded from the study. Out of total 132 patients 78 received carbamazepine analogue as add on therapy to previous anti-epileptic drug (AED) with 58 controls using AED excluding carbamazepine. Study result showed improvement in anger score 7. Catecholamines are primarily produced by adrenal medulla. Specifically dopamine is related to anxiety , anger, depression and aggressive behavior. 32 tonic-clonic epileptic patients were recruited with 29 control subjects in other arm. Urine samples were collected to analyse catecholamine. It was observed that dopamine excretion was decreased in epilepsy patients group 8.

Dopamine inherited gene variation has been identified in two types of epilepsy inherited patients, sleep related seizure and juvenile type of epilepsy 9, 10. Ketogenic diet is therapeutic high fat low carb diet however if sugar metabolism is deranged patients will not be able to digest fat. However there is increasing evidence that gut immunity sensitized with vitamin E, Essential fatty acid, Zinc, fruit ficus carica and gut bacteria Bifidobacterium and Lactobacillus benefits epilepsy patients.

Reference

1) Lakhan R et. al. possible role of of CYP2C9 and CYP2C19single nucleotide polymorphism in drug refractory epilepsy Indian Journal Med 134 Sep 2011; 295-301.

2) Saruwateri J etal, update on genetic polymorphism of drug metabolizing enzyme in antiepileptic drug therapy, Ph armaceuticals, 2010;3, 2709-2732.

3) Tate SK et al., Genetic predictors of maximum dose patient receive during clinical use of anti-epileptic drugs carbamazepine and phenytoin Proc Natl Acad Sci USA 2005;102:5507-5512.

4) Ho PC et al; Influence of CYP2C9 genotype on formation of a hepatotoxic metabolite of valproic acid in human liver microsomes Pharmacogenomics J 2003;3:335-342.

5) Delgado-Escueta AV et al, The nature of aggression during epileptic seizure N. Engh J. Med. 1981 Sep 17; 305 (12): 711-6.

6) Blumer Antidepressant and double antidepressant treatment of affective disorder of epilepsy. Journal of clinical psychiatry 1997.

7) Toledo M. Levels of anger in epilepsy patients treated with escicarbazepine acetate. Acta NeurologicaScand. April 6 2019 doi 10.111/ane. 13099.

8) Kapusteki J et al Secretion of catecholamine in urine of patients with tonic clonic epileptic seizure Wiad Lek 200; 53 (9-10):499-506.

9) Yuri Bozi et al Role of dopamine signaling in epileptogenesis. Front Cell Neurosci 17 Sep 2013 Volume 7, Article 157.

10) Samson Kurt, Dopamine system defect implicated in two forms of epilepsy, neurology today, oct 2 2008, Vol 8, Issue 19 Page No. 32-35.

7. Pathophysiology

In drug discovery point of view research in genomics is restricted due to Cost, Chemical-Biological Safety and Toxicity, Ethical Issues. Genome has a complex Biochemical and physiological function. Though human genome mapping has been fully completed, different response in different patient to same drug is not completely understood. From homeopathic point of view certain group of remedies suffer a specific disease however drug response with respect to remedy constitution will be helpful in safe conduct of trial.

Cystic fibrosis where genetic mutation is related to specific gene, in disease condition cardiovascular disease, Rheumatoid arthritis diabetes schizophrenia network of genes fail to perform correctly. Specific Homeopathic constitution wise allopathic drug prescribing should be considered.

Old concept of regulatory action of pituitary has been replaced by neuro-hormones. These neuro hormones regulate hypothalamus to produce pituitary signals. Recently low dose dexamethasone in inducing psychosis has been elucidated 1, 2, 3. Biological effect of synthetic hormone targets specific cells, however homeostatic neuro hormone disturbance need to be studied in Homeopathic dose. E.g. Thyroid Hormone has role in growth and development of body, Synthetic Hormone has ability to depress anterior pituitary and neuro hormone as hyperthyroidism causes irritability, Hypothyroidism causes sluggishness. Hormone is essential to life Hypo or Hyper function with mild symptoms should not be considered as disease unless it involve vital organ or contribute to major disease (e.g. hypothyroidism link with heart disease) as hypothyroid functioning is still maintaining internal environment. Homeopathic hormonal preparation at 3x and 6x potency should be used to avoid multi-systemic synthetic hormone side effects in every case. Synthetic hormone response differs to different individual; at least one episode of hyperglycaemia is observed in 86% of cases who received high dose steroid 4. Type 2 DM and Steroid Induced DM, Mechanism of insulin resistance is same. Steroid induces 60 to 80 % of insulin resistance depending upon dose and type of steroid 5, 6. Asthma, Arthritis, Infertility, Mental Illness where allopathic research is struggling to minimise systemic side-effects, Homeopathy assures Ideal restoration of health in Individual cases. Personalized Medicine state restoration of health and not treating disease gall stone, uterine fibroid, piles, and haemorrhoids, urinary stone can be treated without operation. Government initiative to register cases through disease registry or Insurance data has major role to play in health service policy.

Because health is basic need and disease treatment is pharmaceutical personal business.

Study literature from allopathic background gives us a clue about pharmacological activity of salbutamol, induces rise in insulin secretion and also has role in heart function 13, 14. Fire cortisol is natural hormone water group of remedies benefits steroid induced diabetes cases specifically acting on sympathetic nervous systemplatina, phosphorus. Steroid Induced DM studies explores light on Drug induced hyperglycemias and Impaired insulin secretion or destruction of Pancreatic beta cell, If Potentised Dexamethasone acts rapidly and controls blood sugar level permanently, It will be helpful to understand patho-physiology of steroid induced diabetes 7. The GC induced diabetes is common cause, found orally administered steroid induced diabetes in 2 % of new cases 8. Glycemic control worsens in DM Cases, However in patients without DM Hyperglycaemia may be transient or permanent 9, 10.

In few studies Decrease Insulin Secretion and Sensitivity is stated as responsible factor for Hyperglycaemia induced by GC, Incidence of GC reported from 2% to 50% 11, 12.

5.2 million cases a year of medical negligence are recorded in India (Medical Errors, Mishap from Medication, Hospital acquired infections, Blood clot developed in leg veins, Incorrect prescription, wrong dose, wrong medication wrong patient, wrong surgery, wrong time to wrong drug) 15.

Similarly long term use of synthetic agonist is linked with serious adverse event and patient death 16. Prior treatment with natural Medicine is must for long term use of

Synthetic hormone, Sympathetic Agonist, Synthetic Neurotransmitter use as protein transport, discrepancy in regulation of hormone cannot evaluated in every case.

Homeopathy is personalized medicine; its mission is restoration of health not treating disease. Awareness and promotion from government will decrease incidence of various fatal diseases HIV, Cancer, Tuberculosis, Diabetes, HTN. Physician Negligence to treat disease is different, Patient negligence and unawareness about Homeopathy need government promotion.

There is correlation between increased negligence cases and Increase in Lab Investigation, unnecessary X-Ray, CT scan, Surgery, and Increase number of drug per prescription. The need of an hour is to realize cost of healthcare can only be controlled by natural medicine.

Acute Hemorrhagic Stroke

Stroke is second leading cause of death and forth leading cause of disability worldwide in 2007 where as 5.7 million deaths estimated in 2005, and 87% of these deaths were in low-income and middle-income countries. (Strong, et.al.2007). In 2001out of all deaths CVA accounted for 5.5 million i.e.9.6 % deaths worldwide. It was estimated that two third of deaths occurred in developing country and patients of less than 70 years of age were 40%. On other hand millions of stroke survivors lead to prolonged disability and restriction in activity need support of other person to survive. In developed countries where approach to the patient with acute stroke has radically changed in last decade with ability to minimise the patient neurological injury with improvement in patient outcome (Mower, et al 1997).

Potential therapeutic benefits can be achieved by managing emergency hypertension in CVA within 48 hours; Easily titratable agents are preferred for these patients to avoid further complications (Sanders, et al 2000) (Shayne, 2003).

CVA Causes interruption of blood flow to brain, damaging brain cells results in impaired body function. In case of Ischemic event cerebral artery is occluded by thrombus or embolism. Whereas, in hemorrhagic CVA cerebral vessel ruptures; blood leaks out in brain tissue. Previous study indicates 80% of CVA cases are ischemic; 20% are hemorrhagic. (National Institute of Health and Clinical Excellence Diagnosis and Initial Management of Acute Stroke, July 2008)

Ischemic CVA can be divided in to two categories embolic and thrombotic. An embolic CVA embolus travels to distant site lodges blocking blood supply to brain, whereas thrombotic CVA is a gradual process of occluding artery by process of atherosclerosis. In thrombotic stroke symptoms appears when patient blood pressure is low systolic < 110 at early morning or at rest.

Systemic vascular resistance which is self-propagating increases by vascular constrictor mechanism leads to hypertensive crisis. Increase in BP severely may result in fibrinoid necrosis of arterioles and endothelial injury (Meredith, et al 1998).

Intravenous preparations require requires continuous blood pressure monitoring unless leads to persistent lowering in BP. whereas oral formulation have limitation as difficult to administered in acute phase of stroke, act irreversibly, and if slow release preparation need to be administered crushing for nasogastric administration destroys its delayed action. Sublingual Depin is frequently used in management of hypertension; which is neither safe nor efficacious should always be avoided. (Sare, et al 2001).

There are four agent that reduce blood pressure in a easily titratable fashion (Require minute to minute adjustment of dose) and do not lead to increase in intracranial pressure these are Labetalol, Esmalolol, Enalaprilat and Nicardipine (Muzzi,et al 1990) (Vaidya,et al 2007) sublingual Nifedipine, are neither safe nor efficacious in this situation; It may result in rapid fall in blood pressure or steal phenomenon. (American Heart Association 2006) There is recent evidence that the trans-dermal administration of nitrates is convenient alternative that does not reduce cerebral blood flow in acute stroke (Spence, 2007). Out of all four antihypertensive Labetalol and Enalaprilat are available but in high cost, Esmalolol is in India's essential drug list, Nicardipine is not available in India.

The early management Hypertension in CVA affects the extent of brain injury and patient suffering's thus we must consider specific treatments (recommended and appropriate antihypertensive agent, thrombolytic agents etc.) from onset of complaint. in stroke patients if blood pressure is not managed efficiently it may leads to the rapid and persistent decrease in blood pressure according to pharmacokinetic and pharmacodynamic profile of drug used and this lowered blood pressure increases risk of ischemic injury and persistent decline in conscious level and death may occur (robinson ,et al 2004).

Review of the Literature with preliminary data

1988: "BEST Trial" 300 patients enrolled in Low dose B-blocker treatment in treatment of acute stroke; Three arms Low dose B-blocker Atenalol, Propranolol, Placebo capsule within 48 hours for three weeks. In patient receiving Beta Blocker more death were recorded as compared to placebo group. After completion of study difference in initial characteristics were observed.

1993: Study, "Should hypertension be treated after acute stroke?".if cerebral perfusion is impaired by decrease in mean blood pressure > 16% below baseline independent of drug used. published in the Archives of Neurology

1995: A retrospective study, 87 patients with Intra cranial haemorrhage with sever hypertension on admission. Higher rate of mortality observed in systolic blood pressure higher than 140mmHg Dandapani et al..

1996-1998 Connecticut Community Hospitals study recommendation was not followed for Blood pressure monitoring. Increased risk of intra cranial haemorrhage was seen in 16% of patients who received t-PA.

2001 13 patients were enrolled in Massachusetts General Hospital study by Dr.Walter Koroshetz and collegues Blood pressure was increased 20% by phenylephrine upto 200 mm Hg.

2002 on the effects of induced hypertension on intracranial pressure and flow velocities in the middle cerebral artery (MCA). In this study, by Schwarz, 2 mm Hg increase in BP, increased MCA velocity by 25 cm/second without increasing ICP. Unfortunately, this was a technical study without any clinical correlate.

2003 (ACCESS) Candesartan used for treatment of hypertension in 342 Ischemic stroke patients; 4-16 mg/day. Trial was terminated because of Candesartan group overwhelming efficacy

2004, Out of 300 patient's 233 patients were found decrease in blood pressure on first day as compared to BP on admission; mean decrease in blood pressure was 8 ± 7 mm Hg in 166 patients who did not received antihypertensive treatment; 36 ± 22 mmHg who received anti hypertensive treatment. Patient were followed till 3 months for outcome; > 20mm Hg reduction in BP were found to be poor out come with higher frequency of neurological deterioration and increase infarct volume.

2005, Antihypertensive agents on the auto-regulatory curve and ACE inhibitors have the most normalizing effect on cerebral auto regulation. So in theory, ACE inhibitors might be useful.

2006 Nitro glycerine paste were used by American Stroke Association (ASA). They also recommended Labetalol and Nicardipine.

2006 29 patients received intravenous Nicardipine. Neurological deterioration was observed in 13% of patients; In 18% of patients haematoma expansion was observed.

2009 A Retrospective study in 100 patients of Intra Cranial Haemorrhage The risk of early neurologic deterioration in ICH begins to increase as SBP falls below 123mmHg (Ohwaki et5 al., 2009).

2010 Negative association in Increase blood pressure and clinical outcome and vice versa. Recent Clinical data suggest moderate reduction in BP cause potential benefits of outcome in Ischemic Stroke.

2010 INTERACT Prospective randomized trial 296 patients were enrolled with Intra Cranial Haemorrhage (ICH) Hematoma growth is attenuated with rapid intensive BP reduction after ICH (Anderson et al., 2010).

2010 A Prospective dose-escalation study was performed in 60 patients of ICH Rapid, intensive BP reduction after ICH was feasible and safe (ATACH investigators, 2010).

2011 A Prospective observational study in 432 patient of (Sub Archenoid Haemorrhage, Intracranial Haemorrhage, Ischemic Stroke) were enrolled; this study found that minimum recorded blood pressure and excessive blood pressure reduction may contribute to mortality after stroke(Mayer et al., 2011)

2010 COSSACS Prospective randomized trial 763 patients were enrolled of Ischemic Stroke and Intracranial Haemorrhage continuing antihypertensive treatment after acute stroke reduced BP; there was no significant difference in 2-week death or dependency (Robinson et al., 2010).

2011 SCAST Prospective randomized trial in 2004 patients of Ischemic stroke, Intracranial Haemorrhage, there was a trend toward increased vascular events and worsening functional outcome in patients started on candesartan after acute stroke (Sandset et al., 2011)

2012 In a study 404 patients were enrolled divided in standard care (BP Lowering Strategy) and treatment group (Intensive care) standard care group have shown accentuation of hematoma growth but no significant difference in death in both group; Antihypertensive agents used were uncommon frusemide,urapidil, phentolamine, and glyceryl trinitrate (Mattia et.al., 2012).

Beta Blockers

It blocks the action of adrenaline with reduction in heart force and decrease in heart rate.

Example: Acebotolol, Esmolol, Labetalol, Metoprolol

Angiotensin Receptor Blocker (ARB):

Angiotensin II binding to blood vessel is blocked by ARB. It helps widen blood vessels which lead to reduce heart activity and volume burden.

Labetalol (Action on ANS receptors)

Labetalol is recommended drug of choice drug by American heart association. It has selective alpha 1 and non-selective beta adrenergic blocking action. Dose dependent blood pressure reduction is observed without affecting heart rate with decrease in renin level. Labetalol inhibit renin activity it also decreases sodium excretion which is subsequent effect of decrease blood pressure. Decrease renin activity down regulates conversion of angiotensinogen to angiotensin 47.

Angiotensin Converting Enzyme

Essential regulator of blood pressure is angiotensin converting enzyme. Angiotensin is peptide hormone. Liver produces angiotensinogen which is broken by renin to produce angiotensinogen I. it does not have specific biological role. It is converted into angiotensinogen II by angiotensin converting enzyme. Angiotensin II causes vascular constriction, stimulates aldosterone production with loss of sodium and potassium ion from kidneys.

Angiotensin converting enzyme inhibitor example:

Captopril, Parindopril, Ramipril, Trandopril

Angiotensin receptor antagonist example:

Losartan, Telmisartan, Olmesartan

Adrenalin is linked to homoeopathic phosphorus group of remedies. Scientific studies states that increase sympathetic activity epinephrine release is found to decrease serum phosphorus level 48.

Similarly food supplement nitric oxide decreases release of renin and inhibit vasopressin secretion 49, 50.

Homoeopathic phosphorus and related group of remedies are essential in hypertension treatment. Increase food article containing phosphorus element is a part of diet therapy to hypertensive patient 51. Nitric oxide constraining food supplements such as water melon and L arginine helps improve heart function by regulatory check on renin angiotensin system.

Reference

1. The hypothalamic–pituitary– adrenal axis and depressive disorder: recent progress. Nihon Shinkei Seishin Yakurigaku Zasshi 32, 203–209. Kunugi, H., Ida, I., Owashi, T., Kimura, M., Inoue, Y., Nakagawa, S., Yabana, T., Urushibara, T., Kanai, R., Aihara, M., Yuuki, N., Otsubo, T., Oshima, A., Kudo, K., Inoue, T., Kitaichi, Y., Shirakawa, O., Isogawa, K., Nagayama, H., Kamijima, K., Nanko, S., Kanba, S., Higuchi, T., Mikuni, M., 2006.

2. Assessment of the dexamethasone/CRH test as a state-dependent marker for hypothalamic–pituitary–adrenal (HPA) axis abnormalities in major depressive episode: a Multicenter Study.

Neuropsychopharmacology 31, 212–220. Kunugi, H., Urushibara, T., Nanko, S., 2004.

3. Combined DEX/CRH test among Japanese patients with major depression. Journal of Psychiatric Research 38, 123–128. Kunzel, H.E., Binder, E.B., Nickel, T., Ising, M., Fuchs, B., Majer, M., Pfennig, A., Ernst, G., Kern, N., Schmid, D.A., Uhr, M., Holsboer, F., Modell, S., 2003. Pharmacological and nonpharmacological factors influencing.

4. High incidence of steroid induced hyperglycaemia in Hospital Fong AC, Cheung NW Diabetes Res Clinical Practice.

5. Glucocorticoid induced hyperglycaemia Clore Jn, Turby Hay L, Endocrine Practice 2009, Jul- Aug 15 (5); 469-74.

6. Glucocorticoid and Cardiovascular Risk factor Strohmayer EA, Krakoff CR. Endocrine Metabolism Clin north AM 2011 june 40(2): 409-17, IX.

7. American Diabetes Association: Diagnosis and classification of diabetes mellitus. Diabetes Care 2012, 35(Suppl 1):S64–71.

8. Gulliford M, Charlton J, Latinovic R: Risk of diabetes associated with prescribed glucocorticoids in a large population. Diabetes Care 2006, 29:2728–29.

9.   Saigí Ullastre I, Pérez PA: Hiperglucemia inducida por glucocorticoides. Semin Fund Esp Reumatol 2011, 12:83–90.

10. Clore JN, Thurby-Hay L: Glucocorticoid-induced hyperglycemia. Endocr Pract 2009, 15:469–74.

11. Gulliford M, Charlton J, Latinovic R: Risk of diabetes associated with prescribed glucocorticoids in a large population. Diabetes Care 2006,29:2728–29.

12. Montori VM, Basu A, Erwin PJ, Velosa JA, Gabriel SE, Kudva YC: Posttransplantation diabetes, A systematic review of the literature.

13. B adreno-receptor stimulation potentiate PKB phosphorelation in rat cardiomyocystes via cAMP and PKA Jorid T Stuenaes, Astrid Bolling & Jorgen Jensen.

14. Some metabolic and hormonal rffect of Salbutamol in man Massara F., Fassio V. Cammani F. et al Acta Diabe lat (1976)13:146.

15. Damyanti Datta Doctors in Doc: Will fuzzy law and frivious cases change the way medicine is practiced in India? India Today Oct 9 2014.

16. Serious adverse event and death associated with treatment using long acting beta agonist. Martinez FD, Arizona respiratory center, University of Arizona College of Medicine Tucson, AZ

1.   Additional reference acute hemorrhagic stroke

1. Andrew R. Haas,& Paul E. Marik, 2006,Current Diagnosis and Management of Hypertensive Emergency, CRITICAL CARE ISSUES FOR THE NEPHROLOGIST, Thomas Jefferson University, Philadelphia, Pennsylvania.

2. Alicia Mattia, 2012, Early intensive blood pressure lowering in intracerebral hemorrhage attenuated hematoma growth over 72 hours, but the effect on clinical outcome remains unclear, Evidence Based Neurology Group University of Western Ontario.

3. Bhuvaneswari K. Dandapani, MD; Shuichi Suzuki, MD; Roger E. Kelley, MD; Yolanda Reyes-Iglesias, MD; Robert C. Duncan, PhD.,1989, Relation Between Blood Pressure and Outcome in Intracerebral Hemorrhage, Department of Neurology, Miami.

4. B. Sanders,2003, Stroke Management, Nervous System Disorder, America.

5. Chirag K. Vaidya, MDJason R. Ouellette, MD, 2007, Hypertensive Urgency and Emergency, Turner White Communications, Wayne.

6. Craig S. Anderson et.al.,2012, The Intensive Blood Pressure Reduction in Acute Cerebral Haemorrhage(INTERACT), University of Missouri—Columbia.

7. D H BARER,& J M CRUICKSHANK, S B EBRAHIM, J R A MITCHELL,1988, "Low dose blockade in acute stroke "BEST" trial", British

Medical Journal Volume 296, Clinical Research Alderley Park, Macclesfield.

8. Donald A. Muzzi, MD, Susan Black, MD, Thomas J. Losasso, MD, and Roy F. Cucchiara, MD 1990 Labetalol and Esmolol in the Control of Hypertension After Intracranial Surgery, Department of Anesthesiology,Mayo Clinic, 200 First Street, S.W., Rochester, MN.

9. Erin M. Grise and Opeolu Adeoye,2012, Blood pressure control for acute ischemic and hemorrhagic stroke, Department of Emergency Medicine and Division of Neurocritical Care, Department of Neurosurgery, University of Cincinnati College of Medicine, Cincinnati, Ohio, USA.

10. Gillian M Sare, MRCP(UK); Chamila Geeganage, MSc; Philip M W Bath, FRCP,2006, Stroke Trials Unit, Institute of Neuroscience, , UK High blood pressure in acute stroke: to treat or not to treat?, University of Nottingham.

11. G. Boysen, H. Christensen,2003, Management of Hyperglycemias & Hypertension in Acute Stroke Dept. of Neurology, Bispebjerg Hospital, University of Copenhagen.

12. Guy Rordorf, MD.et.al., Pharmacological Elevation of Blood Pressure in Acute Stroke Clinical Effects and Safety, Copyright © 1997 by American Heart Association Boston.

13. Joachim Schrader,et.al., June 19, 2003, The ACCESS Study: Evaluation of Acute Candesartan Cilexetil Therapy in Stroke survivors Stroke; America.

14. J Potter, A Mistri, F Brodie, J Chernova,E Wilson, C Jagger, M James, G Ford and T Robinson, 2009, Controlling Hypertension and Hypotension Immediately Post Stroke (CHHIPS) – a randomised controlled trial,Health and Technoogy Assessment, UK

15. J. David Spence, MD, 2007 Current Treatment Options in Cardiovascular Medicine New Treatment Options for Hypertension During Acute Ischemic or Hemorrhagic Stroke,America.

16. Karsten Overgaard, Peter Sandercock and Nils-Gunnar Wahlgren Philip Bath, Gudren Boysen, Geoff Donnan, Markku Kaste, Kennedy R Lees, Tom Olsen, 2001, Hypertension in Acute Stroke: What to Do?,America.

17. 2008,Diagnosis and management of acute Stroke and Ischemic attack, National Institute of Health and Clinical Excellence Diagnosis and Initial Management of Acute Stroke Understanding NICE guidance', High Holborn London.

18. Lisk DR,& (Grotta JC, Lamki LM, Tran HD, Taylor JW, Molony DA, Barron), 1993 Aug, Should hypertension be treated after acute stroke? A randomized controlled trial using single photon emission computed tomography, BJ Archives of

Neurology;50(8):855-62, University of Texas Health Science Center, Houston.

19. Mower, Donna M, 1997 Brain attack: Treating acute ischemic CVA, Neurosurgical Clinical Nurse Specialist Medical Center of Delaware, Copyright Springhouse Corporation Mar, Delaware.

20. Ms Deborah, Leibbrandt, 1998, RNS Hospital, Sydney Literature review of Antihypertensive treatment; NSW Therapeutic Assessment Group Inc and NSW Health, MSD Formerly RNS Hospital, Sydney.

21. May 8, 2006, Hypertension in Acute Stroke: A Better Arrow for the Quiver Hypertension is published by the American Heart Association. Hypertension, originally published online.

22. N.B.S. Lozac, 2010,Central Role of Voltage Gated Calcium Channels and Intercellular Calcium Homeostasis in Autism, published, Clinical findings in autism and relevance of dysfunctional calcium signalling.

23. Paul E. Marik and Joseph Varon, 2007, Hypertensive Crises: Challenges and Management, Official publication on American college of Chest Physician,America.

24. Philip H. Shayne, MD, 2003,Severely Increased Blood Pressure in the Emergency Department, CARDIOLOGY/STATE OF THE ART, Atlanta.

25. Qureshi AI, et al., 2006, "Treatment of acute hypertension in patients with intracerebral hemorrhage using American Heart Association guidelines", Crit Care Med, University of Medicine and Dentistry of New Jersey.

26. Thompson G. Robinson, John F. Potter, 2004, Age and Ageing Vol. 33 No. 1British Geriatrics Society, Leicester.

27. Thomas Truelsen, Stephen Begg , Colin Mathers,2000, The global burden of cerebrovascular disease, Non-communicable Diseases and Mental Health Cluster, WHO Geneva .

28. Niaz Ahmed, Per Näsman and Nils Gunnar Wahlgren, 2000, Effect of Intravenous Nimodipine on Blood Pressure and Outcome After Acute Stroke (INWEST), Copyright © 2000 American Heart Association, Inc. All rights reserved, Greenville Avenue, Dallas, TX.

29. Szocs Ildiko1, & Szatmari Szabolcs, Anamaria Jurcau, Aurel Simion, Edith Sisak,Cheryl Renton, Sharon Ellender, Philip Bath, 20012, THE EFFICACY OF NITRIC OXIDE IN STROKE (ENOS) TRIAL, ROMANIAN JOURNAL OF NEUROLOGY, UK.

30. Strandgaard S, Olesen J, Skinh~jE & Lassen N A. Autoregulation of brain, OCTOBER 5, 1987, circulation in severe arterial hypertension. This

Week's Citation Classic, Department of Neurology, Copenhagen University, Denmark.

31. Salvetti et al Effect of increasing dose of labetalol on blood pressure plasma renin activity and aldosterone in hypertensive patients. Clinical stu London 1070 Dec 57: Supp 5 4015-4045.

32. Jean jaques B et al Epinephrine is hypophosphetemic hormone in man. Journal of clinical investigationVolume 71 March 1983 572-578.

33. Reid IA Role of nitric oxide in regulation of renin and vasopressin secretion. Front Endocrinol 1994 Dec 15 (4) 351-83.

34. Chiu YJ et al Effect of blockage of nitric oxide synthesis on renin secretion in human subjects Clin Exp, Hypertension 1999 Oct 21 (7) 1111-27.

35. Paul Elliot et al Dietary Phosphorus and Blood pressure Internation study of Macro and micro Nutrients and Blood Pressure. Hypertension 2008; 51;669-67

8. Nano - Medicine

Late 19th century was evident of intelligent biomaterial; which has changed researcher's perspective towards science and technology. This intelligent biomaterial are envisioned to have huge impact on healthcare from sequential signaling of biomedical molecule, mimicking natural gene, an effective drug carrier, to high resolution diagnostic tool. Dec.29 1959 Fore father of Nanotech science Finman said at American physical society that there is plenty of space at bottom. A highly specific molecule that can act at cellular and gene level 1.

From drug discovery aspect many of new chemical entity (NCE) fail to reach therapeutic potential due to PK/PD profile. Nanotechnology has changed the focus of drug discovery form chemical evaluation to structure of proteins in signaling pathways and development of chemical antibody. It has offered major success in view of biodegradability, biocompability effect of pH and temperature to targeted drug delivery. As we know toxicity profile of a drug depends on absorption, distribution, and metabolism. Nanotech science has excellent therapeutic response in targeted drug delivery. This biomaterial have offered powerful tool to identify diseases early as at the level of single cell of cancer.

Western world is aware of dramatic potential of nano-projects; which has its limitation in financial investments; with major challenge of transforming nano science to commercial pharmaceutical product.

Example of Nanotechnology drug used in Chemotherapy,

1.   Molecule nano generator

2.   Cyclic peptide.

Molecular nano generator is a case of molecules attached with specific antibody surrounding to a radioactive atom. Antibody helps molecular mono-generator to reach inside of cancer cell; the case then break out to release radiation resulting in death of cancer cell. Cyclic peptide has a specific amino acid group which lodge itself in bacterial cell wall disrupts which result in cell death. Both of these drugs have overcome side effect of system generalized drug administration. Nano-architecture have focused on chemical structure of cell and signaling protein molecules. Target validation and identification of protein drug modification.

Investment community have seen growth of nano product in disease detection, drug delivery, molecular targeting to radiofrequency cure of diseases in past decade. Business expert have to look into scientific discipline of nanotechnology from initiation to marketization of a product.

Types of Nano molecule

1. Quantum dots

2. Nanoparticles

3. Nano shells

4. Nanotubes

5. Nano devices

Quantum dots

It has potential to emit light under UV light influence. Wavelength of color depends on the particle size. Researchers have designed nano particle with ability to find defective gene according to size of defective gene. This will help in Identifying defective gene. It also seems to eliminate need of biopsy in future

Nanoparticles

These particles allow identification of affected tissue in disease. Its physical property has allowed transfer of toxic chemical entity inside of cell especially in case of cancer therapeutics. It has been used in medicine as contrast media as diagnostic tool.

Nano shells

Nanoparticles are already developed with antibody which attaches to nano shells which has affinity toward cancer cells. These nano shells have property to generate intense heat on absorption of infrared light; heat generated by molecule is lethal to cells.

Nanotubes

Nanotubes are carbon rods sized around half the molecule of gene .Nano tubules can identify shape of DNA, Identifying tag for important mutations associated with cancer.

Nano device

Nano-architecture aim to design device which will identify precancerous stage, Deliver to target site, to assess effectiveness of treatment 2.

Financing in Nano project

Nanotechnology from lab to market approval is long process due to regulatory evaluation. Though it seems to be bright future market it has to go through a long process from being innovation to complete market product. This makes whole process expensive making investor reluctant to invest in big projects. Clinical scenario is of prime importance as it demonstrates market potential to attract investors. Eg. There are 20 million patients of kidney failure in US which is expected to double in next decade. A drug marketed by Wyeth and Elan immunosuppressant in kidney transplant has proven market returns with enhanced bioavailability and convenient dosage. Nano cure has already yielded opportunities in diagnostic and cancer cure. Company tie ups and government encouragement will play major role in financing of future projects in Europe and US. With improvement in life expectancy and reduction in mortality below 5 years of age, assured drug delivery part, field of oncology and diagnostics nano technology has promising growth. In Developing countries Increase in lifestyle diseases on back ground of communicable diseases will surely affect impact of future intelligent biomaterial 3.

Market Perspective

There is great disparity between diseases in developed and developing countries. In area of communicable disease 80% of HIV deaths and 90% malaria deaths occurred in sub-Sahara African countries. On the other hand lifestyle diseases are more prominent in developed world. Stringent regulatory evaluation for developing a nano- product for safe use in public at clinic setup increases cost. Till now commercial driver for nano market are oncology product 38% of market share in nanotechnology based product. Nano medicine has potential to cross blood brain barrier; so we may expect chemotherapeutic drug acting on brain tumor cells. Another field is cardiovascular diseases with large patient pool. From regenerative neuropathy to nano cosmetics will attribute high growth rate in upcoming future. Tech novist analyst reported nano market will grow at compound annual growth rate (CAGR) 12.5% between 2011 -2016. CNS product CAGR will be 16% between same years. Anticancer products will grow at CAGR of 10%.

Another survey of BBC market research which was prepared on basis of established market of nano medicine. Report was prepared after discussion with 60 leading companies with 5 year sale forecast. Nano drug have shown increase in patent filing in USA 53%, Europe 25 % and Asia 12%. Till 2014 Anticancer drug will achieve largest growth in market segment. Nano market was worth

53 billion in 2009 is projected to grow at CAGR of 13.5% surpassing 100 billion in 2014.

A study of national burro economic research in US reported that there is decrease in funding from 40% to 24% of US share in R&D investment. Competitive position appears to be strong with most nano technology related inventions.

REFERENCES

[1] John F. Sargent Jr. December 16, 2013.,Nanotechnology: A Policy Primer,Specialist in Science and Technology Policy.

[2] Renzo Tomellini et.al, Sep 2005. European technology and platform on Nanomedicinenano, technology for Health.

[3] MoustafaMoustafa Mohamed Ahmed An Overview of NanomedicineProfessor Biophysics and Medical Physics Department, Medical Research Institute, Alexandria University

9. ANS Specific Remedy

Homoeopathic research: past, present and future

In era of target specific drug discovery, Homoeopathy is struggling to construct quality evidence and expedient research method. Multiple outcomes are being studied with potential benefits in Homoeopathic treatments. Single subject trial is an important research area aims to identify individual disease susceptibility. Autonomic Nervous System (ANS) selective remedy response proves law of simplex [1]. Doctrine of drug dynamisation is addressed by identifying Nano particles in Homoeopathic high potency dilution [2]. Nature maintains and regulates ANS activity, Homoeopathic natural principle need to be validated on scientific instruments [3,4]. Homeostasis is regulated by ANS; remedy relationship will give new dimension to Homoeopathy

Homoeopathy is well established system of medicine. The fight over ideal cure and animosity in quality research evidence continues after 222 years. Master Dr. Samuel Hahnemann discovered law of similia and drug potentization. Previously described in Ayurveda, sookshma aushadhi chikitsa - micro fined medicinal particles processed to improve medicinal efficacy. Master Dr. Samuel Hahnemann, ascribed usefulness of cinchona bark in malaria treatment., work was based on natural principles which are exactly opposite to contraria contrariis curantur. He soon faced criticism, with success in treatment of community diseases. At that time drug companies realized the ease of Homoeopathic system of medicine, which leads to rejection of homeopaths in the medical communities. Homoeopathic practitioners faced this challenge with setting up their own institutions. Dr. Hahnemann students got recognition all over the world with integrated, coherent philosophical principles. Dr. James Tyler Kent briefly introduced mental and emotional elements in his book Kent lectures. He contributed in diverse area of research, drug proving, repertory, remedy relationship. Dr. Constantine Hearing introduced Lachesis remedy and Hearings law of cure. Dr. Phillip Bailey explained virtual emotional disturbance and intellectual aspect of constitutional remedies.

The Homoeopathy popularity dropped since major trial shows ambiguous result. Clinical trial is a tool calibrating two system of medicine acting at different levels. Homoeopathic principal similia similibus curantur is exactly opposite of contraria contrariis curantu. Homoeopathic science has two fold existence environmental effect on body and remedy relationship (fig.2). There is further scope to expand logical prescription from environmental changes.

Ever since Master Dr Samuel Hahnemann proved drug Cinchona deliberate evaluation of physical sign and symptoms, psychological profile has been elucidated by proving on healthy individuals; he stated principle of Similia Similibus Curantur. Constitutional remedies; covers variety of physical and mental symptoms. Many issues arise when planning clinical trial. Homoeopathy has its own drug proving method, doctrine of drug proving is assessment of safety and efficacy monitoring of serious adverse event, efficacy: action of remedy physical and mental symptoms by proving it in low doses in healthy human being. Homoeopathy is practiced science 222 years; Individualized medicine with study of disease susceptibility. 5.2 million cases a year of medical negligence are recorded in India medical errors, mishap from medication, hospital acquired infections, incorrect prescription [5].

A cohort of similar sign and symptoms is convenient for Homoeopathic trial. The medical analyzer to record ANS response, Homoeopathic drug proving pertinent evidence is needed to be checked with pragmatic scientific device readings. More over an individual element constitution will give assertive response on medical analyzer (fig.3).

Other Efficacy Parameters

Plasma Hormonal Levels

Improvement in quality of life

Homoeopathic Aggravation

Pre and Post treatment ECG, Urine analysis, Cardiovascular Parameters etc.

The concept of Psora Syphylis and Psychosis is closely aligned with sugar, protein and fat metabolism (fig. 4).

Psora-Sugar Metabolism

Psychosis- Protein Metabolism

Syphilis- Fat Metabolism

Bread agg. (Complex carbohydrate): Ant-c., bar-c., Bry., carb-an., caust., chin., clem., coff., crot-h., crot-t., kali-c., merc., nat-m., nit-ac., nux-v., olnd., ph-ac., phos., Puls., ran-s., rhus-t., ruta., sars., sec., sep., staph., sul-ac., sulph., teucr., zinc., zing.

Meat (Animal source protein): Carb-an., caust., colch., cupr., ferr., kali-bi., lyss., mag-c., mag-m., merc., ptel., puls., ruta., sil., staph., sulph., ter.

Fat agg: Acon., ant-c., ant-t., ars., asaf., bell., bry., carb-an., carb-s., Carb-v., caust., chin., colch., Cycl., dros., eupho., ferr-ar., ferr-m., Ferr., hell., hep., ip., kali-ar., kali-c., kali-chl., kali-n., mag-c., mag-m., meny., merc-c., merc., nat-a., nat-c., nat-m., nat-p., nit-ac., nux-v., phos., ptel., Puls., rob., ruta., sep., sil., spong., staph., sulph., Tarax., thuj., verat.

(Kent Repertory > Generalities Sec > Page Number 1362
22)

Review of Literature

Dr. Akalpita Paranjpe former BARC scientist has
confirmed the action of plant and mineral kingdom on
ANS. Variable dose of 200 to 1M Homoeopathic potency
was first time validated with scientific experiment. Heart
rate and blood flow was studied with Sulphur, Phosphorus,
Gelsemium, Pulsatilla and Aconite[1]. Homoeopathic
principles are needed to be tested with scientific
instruments to record environmental effect on ANS.

Association between physiological patterns and tridosha
theory, prakriti parikshan and Metabolism in Ayurveda is
well explained with knowledge of complex physiological
reactions [7].

Team of IIT Bombay Scientist has proved that extreme
Homoeopathic dilution of gold, copper and iron retains
Nano particles [2].

Homoeopathic medicine retains the property of crude
substance characterized by nuclear magnetic resonance,
this study states that Homoeopathic dilution cannot be
considered as placebo and there is an evidence gelsemium,
copper memory in dilution which can be recorded [8].

Neuronal circuit of autonomic nervous system is responsible for regulation of physiological state in dynamically changing post postpartum environment [9].

Immune response in target cells or lymphoid tissue is a result of diverse signaling pathway which includes neurotransmitter, neuro modulator, and cholinergic anti-inflammatory and adrenergic receptors. Autonomic nervous system regulatory link between immunity associated changes with respect to natural diseases [10].

A study published in Indian Journal of Research in Homoeopathy states effect of Homoeopathic medicine in behavior disorder in autistic children, there was significant improvement in neuropsychological symptoms such as hyperactivity reduction, communication difficulty, sensory impairment and behavioral dysfunction [11].

Factors affecting selection of dose

Disease complexity

Disease severity

Duration of disease development

Major constitutional remedies

Autonomic Nervous System

Disease Complexity

A person with deranged sugar metabolism is more susceptible for natural diseases. Essential response of immunity development can be achieved with small doses. Approximate quality of response to complex protein and fat metabolism can be improved with high doses.

Disease severity

To trace response of remedy low dose is essential, however a physician is assured about totality of symptoms higher scale of potencies will give finer result. ANS affecting minor changes in physiological reaction require low doses example sneezing, bloating. High potencies are more simplified form of crude drug with precise action. Example: Constipation ~ lycopodium 30C, Renal Stone ~ lycopodium 1M.

Duration of disease development

Duration of disease development correlated with its pathophysiological aspect or degree of metabolism disorder. More convenient high dose is required for long term adaptation.

Major constitutional remedies

As compared to plant and animal kingdom similar threshold therapeutic result can be obtained with low dose of mineral kingdom. High dose of animal and plant kingdom gives favorable result.

Animal Kingdom $\geq$ Plant Kingdom $\geq$ mineral Kingdom

Autonomic nervous system

Symptoms related to sympathetic nervous system responds to low dose and parasympathetic nervous system need higher dose

Immunity

Channelizing immunity in right direction with food is a progressive natural phenomenon in every individual. Immunity development is succession of various challenges (threat) posed by environment and bodily response to it is new path of investigation. The term immunity is in state of chaos for modern medicine physicians.

On basis of observation immunity is evolving state of disease metabolism a double edge sword. One good example is evolving state of protein metabolism and tuberculosis. There are ever increasing number of invention in modern medicine, however scientific treatment rationale for genetic disorder is at the initial stage of drug development. A study published in medical principles and practice journal 2014, conducted at Mumune hospital turkey investigated metabolism in beta thalassemia minor. Out of 194 patients 92 enrolled in Beta thalassemia group and 102 enrolled in control group without beta thalassemia [12]. The fasting insulin, fasting glucose, HDL and triglycerides were higher in study group.

Other studies have also correlated increased glucose and insulin levels. Bahar et al reported that development of insulin resistance in subjects with beta thalassemia was more frequent due to oxidative stress and hemolysis [13]. Tong et al observed normal glucose tolerance with insulin resistance and high fasting glucose level in thalassemia minor subjects [14]. Increase in HDL and Triglyceride is associated with low LDL [15,16].

There is little recent reliable approach in Ayurveda herbs without complete recovery. There are anecdotal reports on Hydroxyurea and wheatgrass to reduce transfusion dependencies in Thalassemia   cases [17,18].

Homoeopathic clinical studies have shown encouraging results, it is claimed that Homoeopathy offers supportive treatment to thalassemia Homoeopathy address to root cause of diseases, which decrease frequency of blood transfusion required. Immunity sensitization with Homoeopathic medicine is required in suspected cases [19].

Formal knowledge of disease symptoms and dosage dose not restore vital force, more than totality is needed an art to prescribe remedies beyond subjective symptoms.

Psora

The secondary manifestation of psora is more dormant form, susceptible to external stimulus. Anger is causes deviation from healthy state of psora. Anger is not a disease state a state that deranges health [20,21]. Ectoderm is related to itch and protection, associated with skin diseases.

Sycosis

The symptoms are first produced by suppressed emotions, followed by inflammatory process. A state of transition subdued quarrelsome nature into state of jealousy, deceitful behavior.

Syphilis

The miasm states functional and emotional state of distruction. Reserved energy deranged by emotional turmoil. Syphilis is susceptible to action of mercurius and natrum mur.

Psora

Probiotic:Enterococcus Strain

Vitamin:Vit D, Vit B Complex

Other food supplement

Alpha Lipoic Acid, Corn, Fructose, Sucrose, Oryza Sativa, Chondroitin Sulphate, Allium Sativum, Glucosamine Sulphate

Magnesium Sulphate, Methionine, Silybum Marianum, Methylsulphonylmethane (MSM), N Acetyle Cystein

Taurine, Nitric Oxide, Cynara Scolymus, Cruciferous Veg

(Susceptible to: Parasite$\geq$ Bacteria $\geq$ Virus)

(Metabolism: Sugar)

Avoid: Complex Carb, Cookies, Fermented Food, Cake

Sycosis

Probiotic:Bifido Bacterium Strain, Lacto bacillus Strain

Vitamin:Vit C, Vit A

Other food supplements

Essential Fatty Acid, Juglans, Sea Food, Theobroma Cacao, Dark Chocolate, Mushrooms, Zingiber officinale, Apple Cider Vinegar, Cinnamomum Verum, Anans Comosus, Citrus fruit, Honey, Crocus Sativus, Astazanthine, Zin Sulphate (Susceptible to: Parasite$\leq$ Bacteria $\leq$ Virus)

(Derranged Metabolism: (Sugar$\leq$ Protein $\leq$ Fat)

Avoid: Nitric Oxide, Nicotiana Tabacum, Polyunsaturated Fatty Acid

Syphilis

Probiotic

Galacto Oligosacchariade, Lactobacillus Rhamnosus, Lactobacillus Bulgaris, Bifido Bacterium

longum

Vitamin: Vit E

Other food supplements: Plant Protein, Momordia Chariantica, Lens Culinaris Medic, Chromium picolinate, Emblica officinalis, Daucus carota, Citrus aurantifolia, Legumes, Linum usitatissimum, Omega Fatty Acid, Whey protein, Branch chain amino acid, Collagen, Hemp protein, Rice Protein, Zea Mays.Cucurbita pumpkin seed protein, Coffea, Camellia sinensis, Solanum lycopersicum, (Susceptible to: Parasite $\leq$ Bacteria $\geq$ Virus)

(Deranged Metabolism: Sugar $\leq$ Protein)

Avoid: Complex protein grains Triticum, Horedum Vulgare, Animal source of protein

Homeopath gathers totality of signs and symptoms, prescribes according to Homoeopathic philosophy. Randomized controlled trial is based on cases, however deep rooted action of Homoeopathic medicine improves disease susceptibility. We need to address causality factor, cohort associated decrease risk of disease development after Homoeopathic treatment [6]. Rationale using remedy specific to environmental change apprise with remedy relationship (table.1). Autonomy in prescribing should be narrowed to obligated remedy relationship. There is modest clinical evidence between sulphur and lycopodium, the remedy relationship is comprehensive with phytochemical evaluation. The symptom similarity between two remedies is closely aligned with chemical profile of medicine to conceive physiological disturbance in correlated autonomic nervous system.

Integrated data in government organization and hospitals would be interesting study subject, to evaluate role of different constitution in disease pathophysiology. There is further scope to classify individualized subordinate food supplements with nutrition concept depicted in Ayurveda.

Reference

1. Mishra N, Muraleedharan KC, Paranjpe AS, Munta DK, Singh H, Nayak C., An exploratory study on scientific investigations in Homoeopathy using medical analyzer. J Altern Complement

Med. 17(8), 2011, 70510. doi: 10.1089/acm.2010.0....334.

2. Chikramane PS, Suresh AK, Bellare JR, Kane SG. Extreme Homoeopathic dilutions retain starting materials: A nanoparticulate perspective. Elsevier Homoeopathy. 2010 Oct;99(4):231-42. doi: 10.1016/j.homp.2010.05.006.

3. Gladwell VF et. al., The effect of views of nature on autonomic controle,Eur J Appl Physiol 112(9), 2012, 3379-86.

4. Abdullah Alabdulgader et al, Long term study of heart rate variability response to changes in solar and geomagnetic environment. Scientific reports 8, 2018, 2663, doi: 10.1038/s41598-018-20932-x.

5. Damyanti Datta Doctors in Doc: Will fuzzy law and frivious cases change the way medicine is practiced in India? India Today 9, 2014.

6. Lin Y et. al. A prospective cohort study of cigarette smoking and pancreatic cancer in Japan, Cancer Causes Control, 13(3), 2002, 249-54

7. Dey S, Pahwa P, Prakriti and its associations with metabolism, chronic diseases, and genotypes: Possibilities of new born screening and a lifetime of personalized prevention. J Ayurveda Integr Med. 5(1), 2014, 15-24. doi: 10.4103/0975-9476.128848.

8. MichelVan, Wassen Hoven, MartineGoyensMarcHenry,EtienneCapieaux,Phili

ppeDevos, Nuclear Magnetic Resonance characterization of traditional Homoeopathically manufactured copper (Cuprum metallicum) and plant (Gelsemium sempervirens) medicines and controls, Elsevier Homoeopathy, 106(4), 2017, 223-239.

9. Stephen W. Porges1 and Senta A. Furman, The Early Development of the Autonomic Nervous System Provides a Neural Platform for Social Behavior: A Polyvagal Perspective Infant Child Dev. 20(1), 2011, 106–118. doi:10.1002/icd.6.

10. MJ Kenney and CK Ganta, Autonomic Nervous System and Immune System Interactions, Compr Physiol. 4(3), 2014, 1177–1200. doi:10.1002/cphy.c130051.

11. Praful M Barvalia, Piyush M Oza, Amit H Daftary, Vijaya S Patil, Vinita S Agarwal, Ashish R Mehta Effectiveness of Homoeopathic therapeutics in the management of childhood autism disorder indian journal of Research in Homoeopathy 8(3), 2014, 147-159.

12. Sinan Kirim et.al. Is beta thalassemia minor is associated with metabolic disorder, Medical principle and practice karger publishers 23(5), 2014, 421-425.

13. Adele Bahar et. al. relationship between beta-globulin gene carrier state and insulin resistance, Diabetes Metabolic Disorder Journal 11(22), 2012, PMCID PMC 3598161.

14. Peter Tong et. al., C reactive protein and insulin resistance in subjects with Thalassemia minor and a family history of diabetes, diabetes care journal 1480, 2002, 1481.

15. Mario Maioli et. al. plasma lipid in beta thalassemia minor, Atherosclerosis 75(2-3), 1989, 245-8.

16. M Hashemieh et. al. Bangladesh Med. Res. Counc Bull 37(1), 2011, 24-7.

17. Dharma R Chaudhari et. al. Effect of wheat grass therapy on transfusion requirement in Beta Thalassemia major,indian journal of pediatrics Volume 76 April 2009 DOI 10.1007/s12098-009-0004-6.

18. Karimi M, Hydroxyurea in the treatment of Thalassemia intermedia, Hemoglobin 33(1), 2009, S177-82. Doi: 10:3109/0360260903351809.

19. Antara Banerjee, Sudeepa Basu et. al. Can Homoeopathic medicine bring additional benefits to thalassemia patients on hydroxyurea therapy? Encouraging result of Homoeopathic treatment. Evidence based complementary alternative medicine, 7(1), 2010, 129-136.

20. Vitaliano PP et. al glucose and insulin relationship with hassles, anger and hostility in older non diabetic adult Dept of psychiatry and behavioural sciences university of Washington seatle USA, Phychosom Med 1996, 489:99.

21. Vera k Tsenkova, Anger, Adiposity and Glucose control in nondisabetic adult Finding from Midus II, J Behav MedFeb 37(1), 2014, 37-46 PMCID: PMC3652917.

22. J. T. Kent, Kent's Repertory of Homoepathic Material Medica, Sixth American Edition edited and revised by Clara Louise Kent, M.D. Reprint edition. Generalities 2004, 1362.

10. Tables and Figures

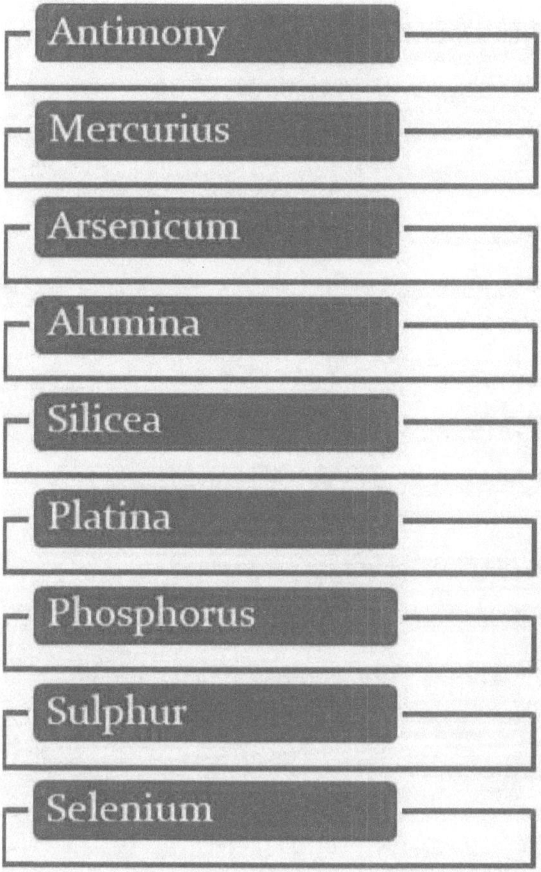

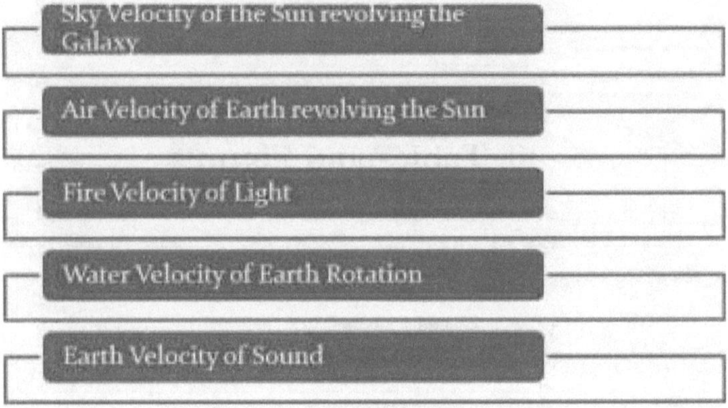

Sky Velocity of the Sun revolving the Galaxy

Air Velocity of Earth revolving the Sun

Fire Velocity of Light

Water Velocity of Earth Rotation

Earth Velocity of Sound

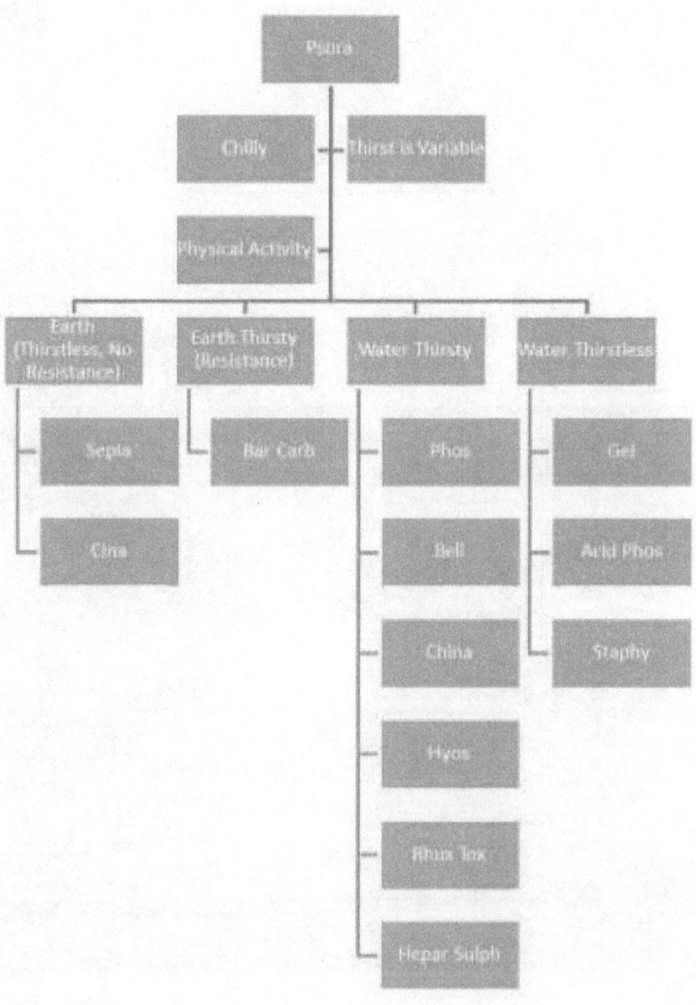

Psora
Chilly
Thirst is Variable
Physical Activity
Earth (Thirstless, No Resistance)
Earth Thirsty (Resistance)
Water Thirsty
Water Thirstless
Sepia
Bar Carb
Phos
Gel
Cina
Bell
Acid Phos
China
Staphy
Hyos
Rhus Tox
Hepar Sulph

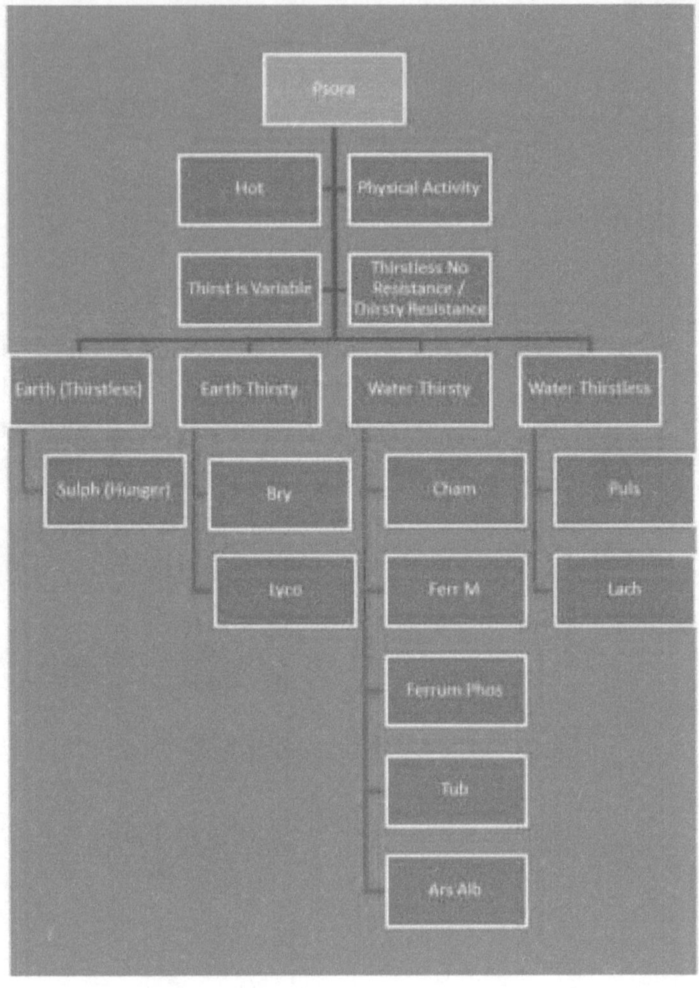

Psora
Hot
Physical Activity
Thirst Is Variable
Thirstless No Resistance / Thirsty Resistance
Earth (Thirstless)
Earth Thirsty
Water Thirsty
Water Thirstless
Sulph (Hunger)
Bry
Cham
Puls
Lyco
Ferr M
Lach
Ferrum Phos
Tub
Ars Alb

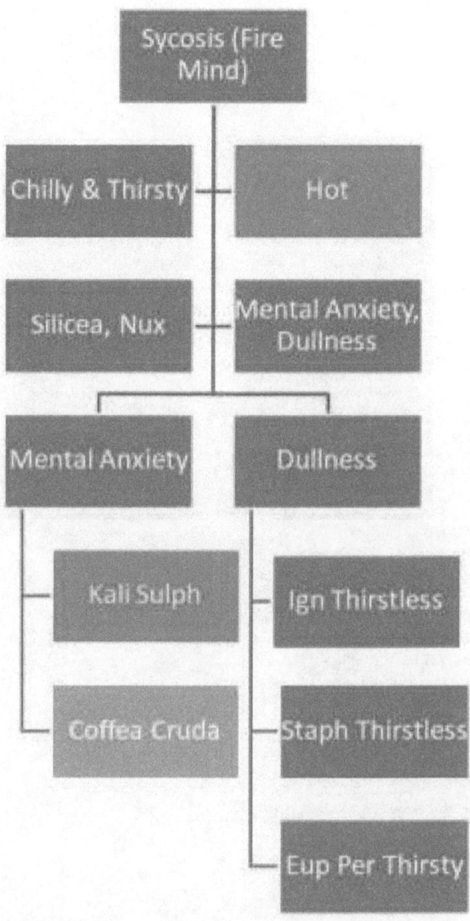

Sycosis (Fire Mind)
Chilly & Thirsty
Hot
Silicea, Nux
Mental Anxiety, Dullness
Mental Anxiety
Dullness
Kali Sulph
Ign Thirstless
Coffea Cruda
Staph Thirstless
Eup Per Thirsty

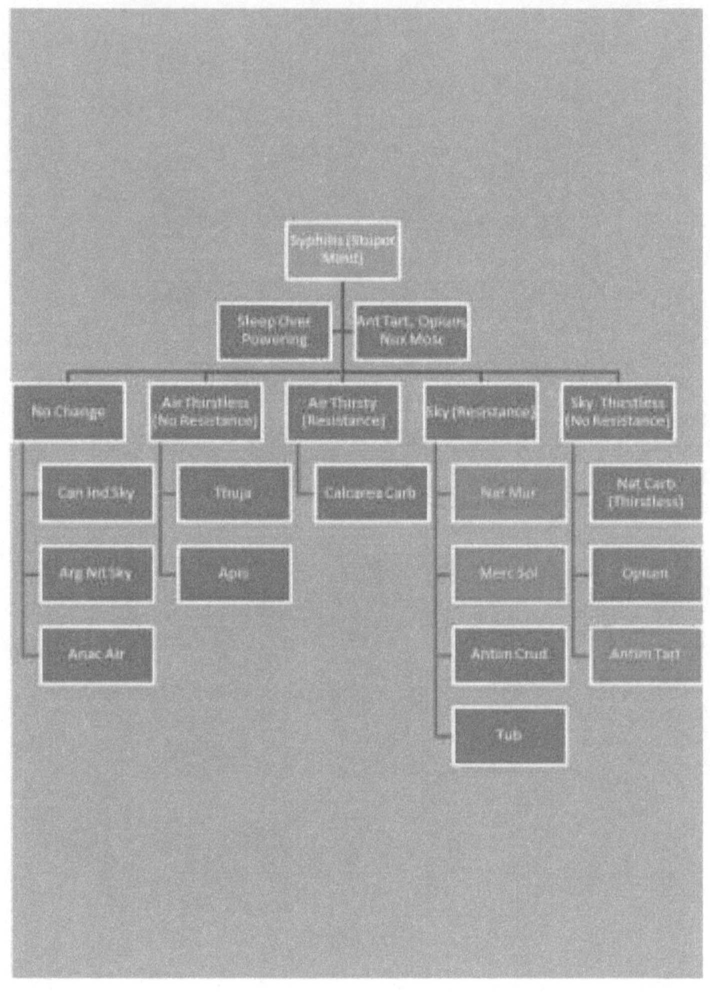

Syphilis (Stupor Mind)
Sleep Over Powering
Ant Tart, Opium, Nux Mosc
No Change
Air Thirstless (No Resistance)
Air Thirsty (Resistance)
Sky (Resistance)
Sky Thirstless (No Resistance)
Can Ind Sky
Thuja
Calcarea Carb
Nat Mur
Nat Carb (Thirstless)
Arg Nit Sky
Apis
Merc Sol
Opium
Anac Air
Antim Crud
Antim Tart
Tub

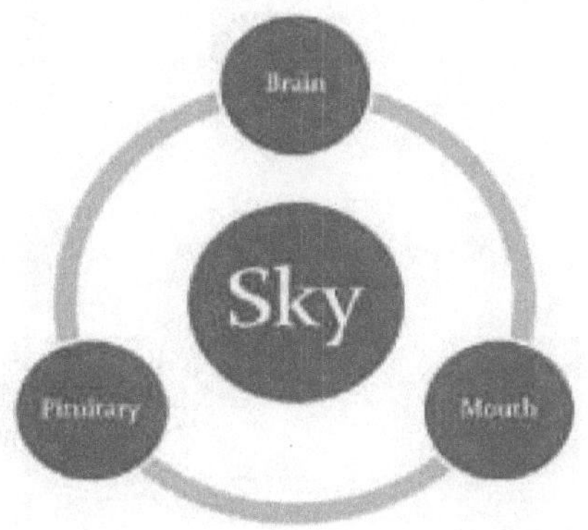

Brain
Sky
Pituitary
Mouth

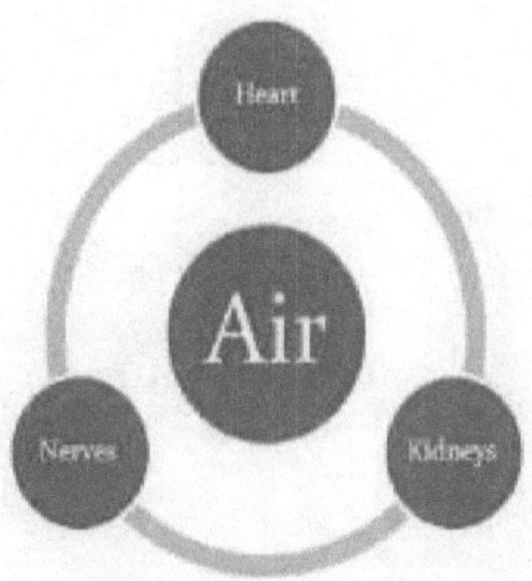

Heart
Air
Nerves
Kidneys

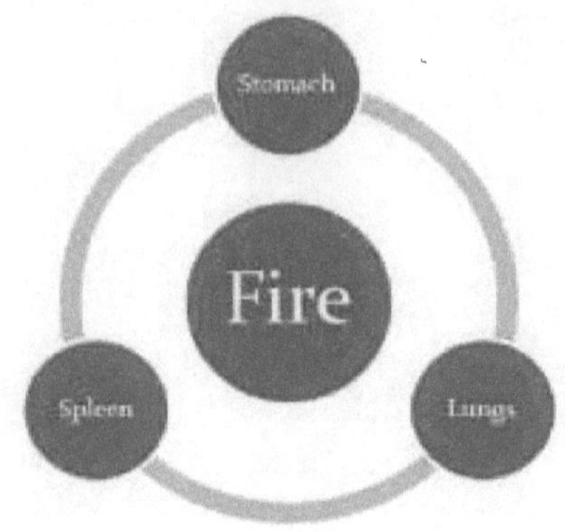

Stomach
Fire
Spleen
Lungs

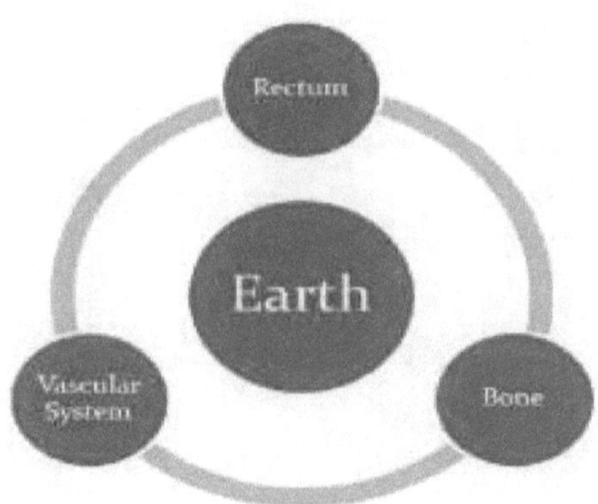

Rectum
Earth
Vascular System
Bone

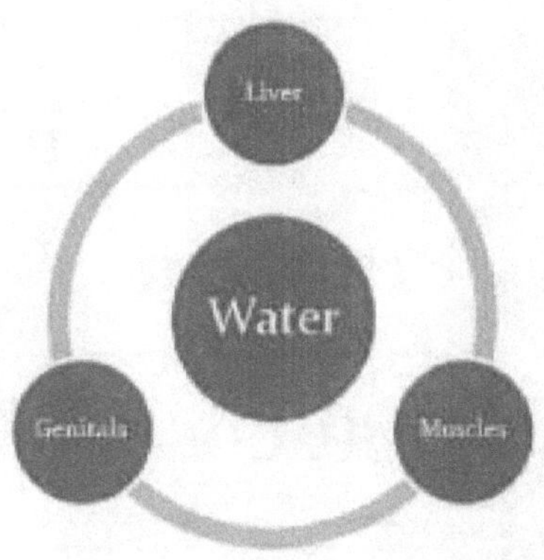

Liver
Water
Genitals
Muscles

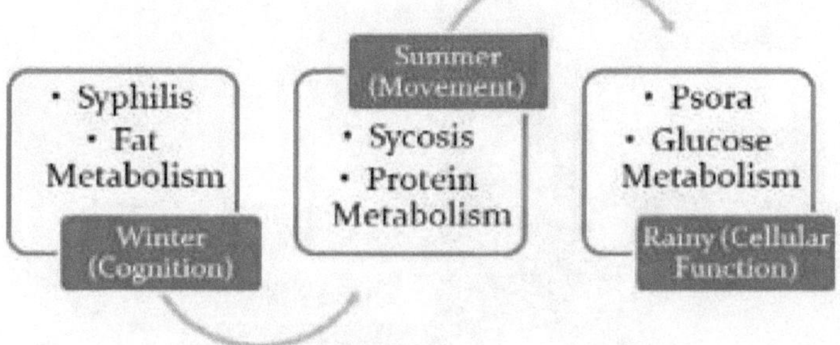

Syphilis
Fat Metabolism
Winter (Cognition)
Summer (Movement)
Sycosis
Protein Metabolism
Psora
Glucose Metabolism
Rainy (Cellular Function)

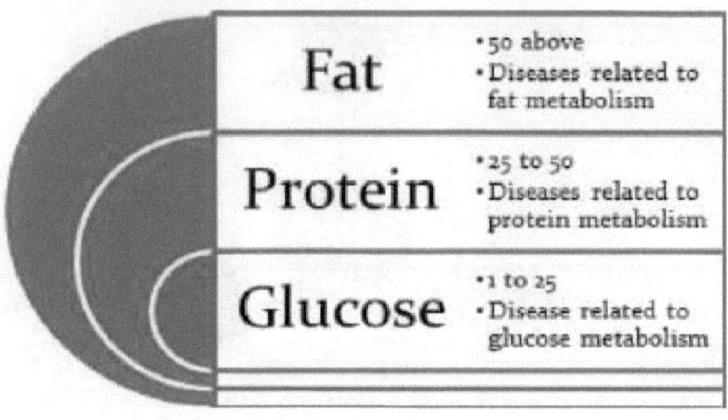

Fat
•50 above
•Diseases related to fat metabolism
Protein
•25 to 50
•Diseases related to protein metabolism
Glucose
•1 to 25
•Disease related to glucose metabolism

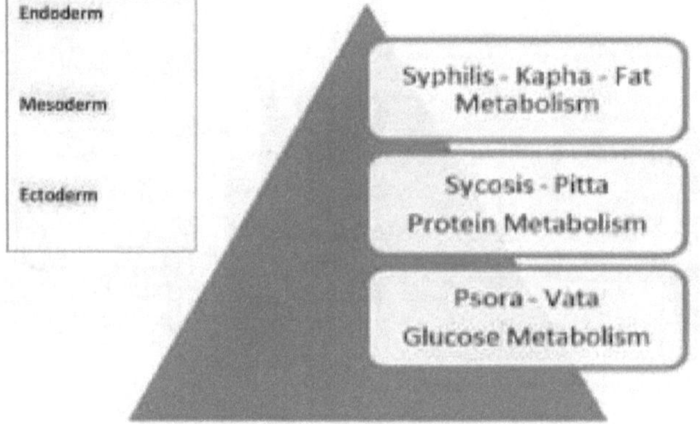

Endoderm
Mesoderm
Ectoderm
Syphilis - Kapha - Fat Metabolism
Sycosis - Pitta Protein Metabolism
Psora - Vata Glucose Metabolism

	Sulphur	Phosphorus	Silica
SNS Element and Animal	Earth-Earth	Water – Earth	Fire Earth
	Earth-Water	Water-Water	Fire Water
	Earth Fire	Water Fire	Fire - Fire
	Earth Air	Water Air	Fire Air
	Earth Sky	Water Sky	Fire Sky
	Sulph	**Phosr**	**Sil**
	Lyco	Bella	Ignatia
	Colchic	Dulca Puls	Spigelia
	All Ce Sab	Tabac Ran B	Gels
	All Sat	Hyos	Strychnin
	Ve Alb Alo	Stram Acon	Nux Vom

	Toad	Snake	Sepia
PSN Bio Salt and plants	Mag Sulph	Fer Sulph	Kali Sulph
	Mag Phos	Fer Phos	Kali Phos
	Mag Flour	Fer Citri	Kali Fluo
	Mag Carb	Fer Ars	Kali Carb
	Mag Mur	Fer Mur	Kali Mur

Elements		
Alumina	Iodine	Bromium
Argentum Metalicum (Pure Silver)	Mercurius (Mercurio)	Mercurius Cynide
Arsenic Oxide	Phosphorus	Stannum Met
Aurum Met	Platina (Platina)	Sulphur
Cuprum Met (Copper)	Plumbum Metalicum (Lead)	Zincum Met
Iron (Ferrum Met)	Selenium	Sulphide
Arsenicum iodatum	Mercurius Dulcis	Antimonium Crudum
	Antimonium Arsenicosum	

Mineral Salt		
Amonium Carb	Nat Carb	Kali Carbonicum
Antim tart	Mag Phos	Magnesia Carbonica
Barium Carb	Kali Brom	Kali Phos
Borax	Ferrum Phos	Kali Sulph
Calcarea Carb	Cal Sulph	Kali Mur
Calcium Carbonate	Cal Fluor	Lithium Carb
Calcarea Phosphorica	Baryta Mur	Mag Muriata
Causticum	Silicea	Nat Phos
Hepar Sulph	Ammonium Mur	Nat Mur
Kali Bichromatum	Amylenum	Ferrum Muriaticum

Ranunculaceae		
Aconite	Hydrastis Canadensis	Clematis Erecta
Actea Recemosa	Pulsatilla	Actea Spicata
Helleborus Niger	Ranunculus Bulbosus	Staphysagria

Liliaceae		
Aloe Socrotina	Trillium Pendulum	Aspargus Officinealis
Allium Cepa	Scilia maritime	Sabadilla
Allium Sativum	Colchicum	Sarsaparilla
Veratrum Viride	Lilium Tigrinum	Veratrum Album

Other			
Carbo Veg (Wood Charcoal)	Graphitis (Black Lead)	Kreosotum (Distillation of Wood)	Sanicula (Natural Spring Water - Children Constitutional Remedy)
Glonine (Nitro Glycerine)	Petroleum (Coal Oil)	Taberenthina (Vegetable Oil)	X-Ray (Radium Bromatum)
	Carbo Animalis (Animal Charcoal)	Fel Tauri	

Other			
Carbo Veg (Wood Charcoal)	Graphitis (Black Lead)	Kreosotum (Distillation of Wood)	Sanicula (Natural Spring Water - Children Constitutional Remedy)
Glonine (Nitro Glycerine)	Petroleum (Coal Oil)	Taberenthina (Vegetable Oil)	X-Ray (Radium Bromatum)
	Carbo Animalis (Animal Charcoal)	Fel Tauri	

Compositae		
Abrotanum	Cina	Millefolium
Calendula	Arnica Montana	Erigeron Canadensis
Chamomilla	Epatorium Perfolatum	Cardus Maranus
Artemesia Vulgaris	Taraxacum	Bellis Perennis

Animal		
Apis Mel (Poison of Honey Bee)	Murex Purpura (Sea Snail)	Mephetis (Liquid Contained in Anal Gland of Skunk)
Cantharis (Spanish Fly Poison)	Vipera (German Viper)	Blatta Orientalis (Cocroch)
Crottalus Horridus (Rattle Snake Poison)	Terentula Hispania (Spanish Spider)	Asteria Rubens (Star Fish)
Lac Caninum (Dogs Milk)	Tarentula Cubensis (Cuban Spider)	Sepia Mallusca (Cuttle Fish)
Lachesis (Surukuku Snake poison)	Naja Tripudins (Cobra Poison)	Bufo (Poison from toad Skin)
Heloderma	Moschus (Musk Deer)	Ambra Grisea (Product fond in belly of sperm Whale)

Nosode		
Medorrhinum	Diptherium (Diptheric Membrane Trituration)	Variolinum (Pus From Small Pox)
Psorinum	Carsinosin	Tubercullinum Bacillinum (Pus from Tubercular Abscess)
Lyssine (Saliva of Rabid Dog)	Antracium (Septicemia, Carbuncle, Felon, Blood Boils)	Syphillinum (Syphilitic Discharge)

Loganaceae		
Nux Vomica	Hoang Nan	Gelsemium
Ignatia	Curare Upas T	Strychnine
	Spigelia	

Solanaceae		
Solanum Nigrum.	Solanum Pseudocapsicum.	Solanum Capense.
Solanum Arrebenta.	Solanum Tuberosum.	Solanum Erythracantum.
Solanum Carolinense.	Solanum Vesicarium Physalis Alkekengi	Solanum Integri.
Solanum Lycopersicum	Solanum Villosum.	Solanum Melongeua.
Solanum Mammosum.	Solanum Xanthocarpum.	Solanum Nodiflorum.
Solanum Malacoxylon.	Solanimum.	Solanum Sodomoeum.
Solanum Oleraceum.	Solaninum Aceticum.	Nicotianum Tabacum.
	Nicotinum	

Papaveraceae	Umbliiferae	Rubiaceae	Coniferae
Chelidonium Majus	Cicuta Verosa	Cinchona Officinalis	Abis Nigra
Opium	Hydrocotyle Asiatica	Ipecacuna	Abies Canadensis
Sanguinaria	Asafoetida	Coffea Cruda	Thuja Occidentalis
Scrophulariaceae	Polygonaceac	Caprifoliaceae	Sabina
Digitalis Purpura	Rheum	Viburnum Ophulus	Trillium Pendulum
Euphrasia	Rumex Crispus	Sambucus Nigra	Sarsaparilla
Droseraceae	Hamamelaceae	Hypericaceae	Gramineae
Drosera Rotundifolia	Hymamelis Verginia	Hypericum	Avena Sativa
Colentirata	Polygalaceae	Passifloraceae	Gentianaceae
Spongia Tosta	Ratanhia	Passiflora Incarnata	Menyanthes Trifoliata
Aslepiadaceae	Labiate	Linae	Simanubaceae
Condurago	Collinsonia Canadensis	Coca (Cocain)	Cedron
Rutaceae	Ramnacede	Verbenaceae	Sapindaceae
Ruta Graveolance	Ceanuthus Ameriuns	Agnus Cactus	Aesculus Hippocastum

Earth	Remedies
Plant Familes Liliaceae Compositae Cactacce	Abrotanum, Aesculus Hispocastanum, Allium Cepa, Aloe Socotrna, Aracardium Orientale, Arnica Montana, Cactus Grandiflorus, Calendula Officina is, Chamomilla,, Cina Maritima, Colchicum Automnale, Eupatorium Perfoliatum,Lilium Tigrinum, Lycopodium Clavatum, Sabadilla, Taraxacum, Cardus Marianus

Water	Remedies
Plant Families Ranunculaceae Solanaceae	Aconitum Nepellus, Berberis Vulgaris, Belladona, Caulophyllum Thalictroides, Dulcamara, Helleborus Niger, Hydrastis Canadensis, Hyoscyamus Niger, Ranunculus Bulbosus, Staphysagria,

Air	Remedies
Plant Families Umblifarae Anacardiaceae Coniferae Papveraceae	Aethuja Cynapium, Anacardium Orientale, Camphora Officinalis, Chellidonium Majus, Conium Maculatum, Nux Moschata, Opium, Sabina, Sanguinaria Canadensis, Sarsaparilla, Synzygium jambolanum, Thuja Occidentalis.

Air	Remedies
Plant Families Umblifarae Anacardiaceae Coniferae Papveraceae	Aethuja Cynapium, Anacardium Orientale, Camphora Officinalis, Chellidonium Majus, Conium Maculatum, Nux Moschata, Opium, Sabina, Sanguinaria Canadensis, Sarsaparilla, Synzygium jambolanum, Thuja Occidentalis.

Sky	Remedies
Plant Families Cucurbitaceae Urticaceae	Bryonia Alba, Cannabis Sativa, Cannabis Indica, Colocynthis, Momordia Chariantica

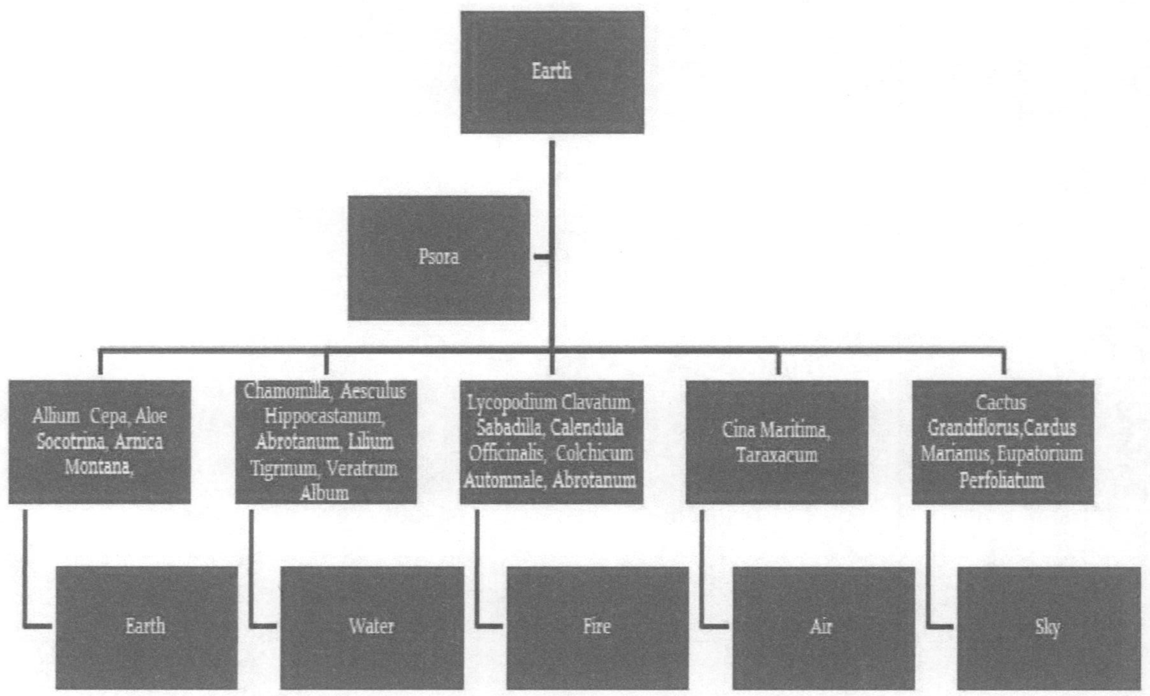

Earth
Psora
Allium Cepa, Aloe Socotrina, Arnica Montana,
Chamomilla, Aesculus Hippocastanum, Abrotanum, Lilium Tigrinum, Veratrum Album
Lycopodium Clavatum, Sabadilla, Calendula Officinalis, Colchicum Automnale, Abrotanum
Cina Maritima, Taraxacum
Cactus Grandiflorus, Cardus Marianus, Eupatorium Perfoliatum
Earth
Water
Fire
Air
Sky

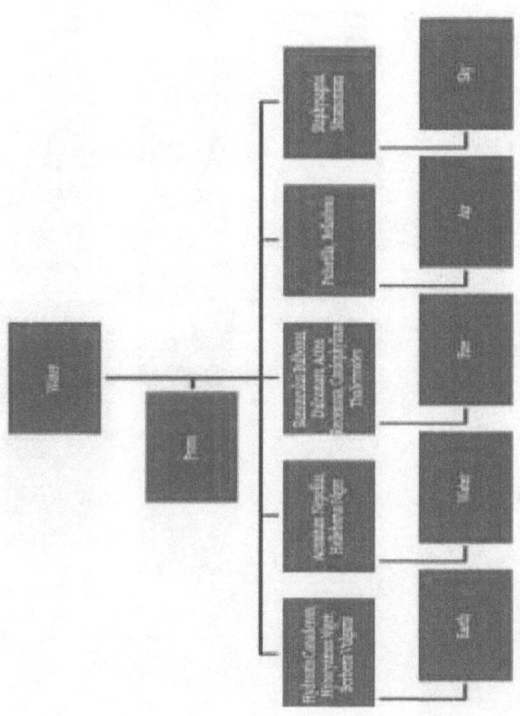

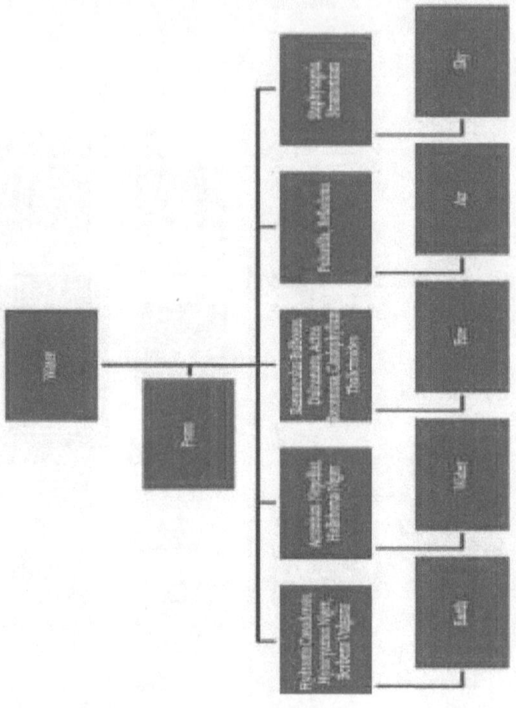

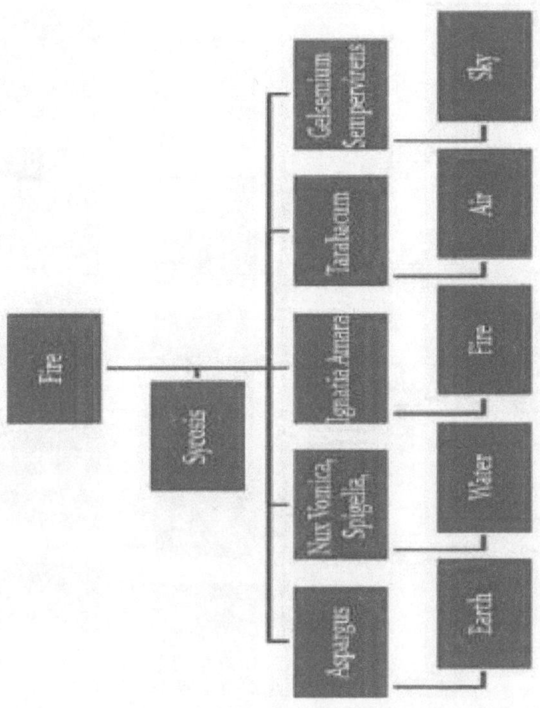

Fire
Sycosis
Gelsemium Sempervirens
Tarahticum
Ignatia Amara
Nux Vomica, Spigelia
Asparagus
Sky
Air
Fire
Water
Earth

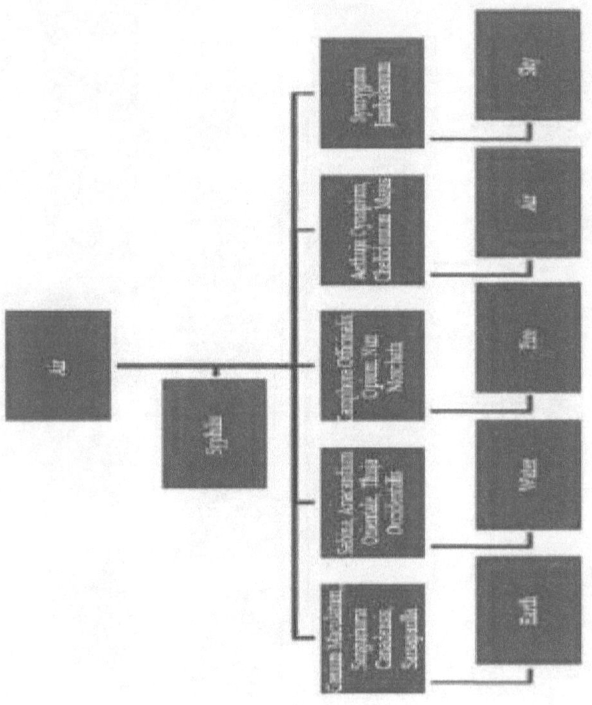

Air
Syphilis
Conium Maculatum, Sanguinaria Canadensis, Sarsaparilla
Sabina, Anacardium Orientale, Thuja Occidentalis
Camphora Officinalis, Opium, Nux Moschata
Aethusa Cynapium, Chelidonium Majus
Syzygium Jambolanum
Earth
Water
Fire
Air
Sky

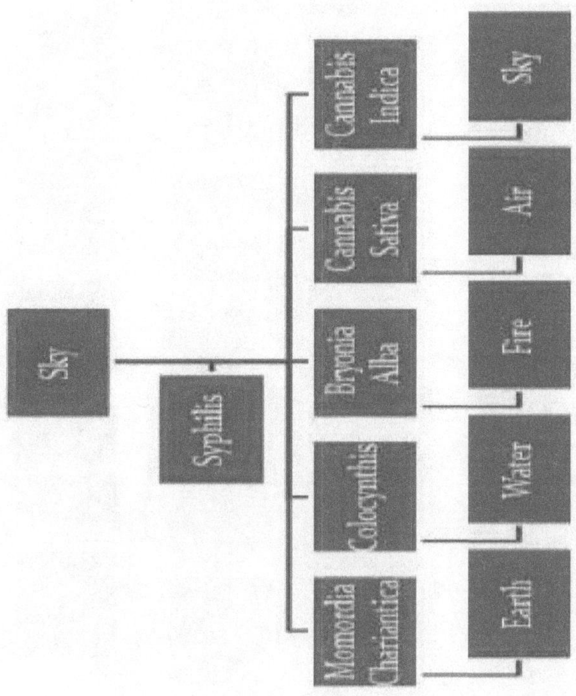
Sky
Syphilis
Cannabis Indica
Cannabis Sativa
Bryonia Alba
Colocynthis
Momordia Chariantica
Sky
Air
Fire
Water
Earth

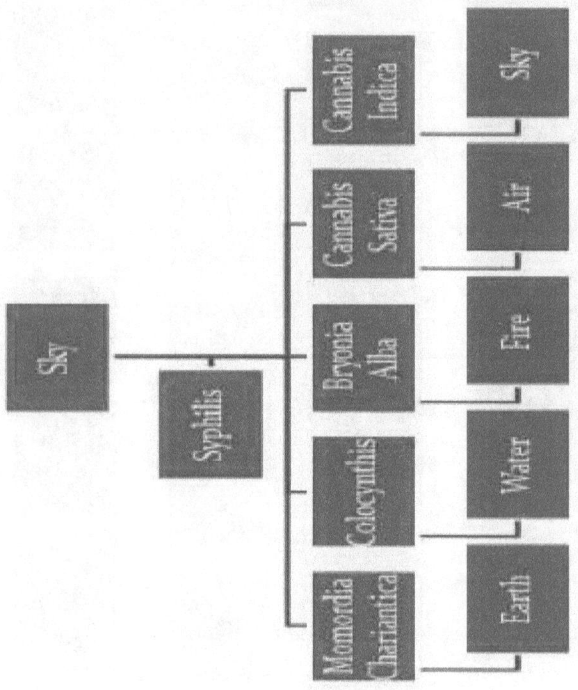

Sky
Syphilis
Cannabis Indica
Cannabis Sativa
Bryonia Alba
Colocynthis
Momordia Charantica
Sky
Air
Fire
Water
Earth

Earth	Water	Fire	Air	Sky
Alpha Lipoic Acid	Z. Jujuba	Plant Protein	Essential Fatty Acid	Essential Fatty Acid
Corn	Soyabean	Momordia Chariantica	Calcium Supplement	Walnut
Fruits – Carb source: Fructose, Sucrose	Spinach	lens culinaris medic	Fish Oil	Sea Food
Rice	Brown Rice	Chromium picolinate	Johns Root	Meat
Papaya	Raisins	Emblica officinalis	Korean Ginseng	Sodium Salt
Chondroitin Sulphate	Cucumber	Daucus carota	Siberian Ginseng	Cocoa
DMSA	Hemp Leaves	Citrus aurantifolia	Fig	Dark Chocolate
Garlic	Tomato	Legumes	Sea Food	Mushrooms
Glucosamine Sulphate	Peanut	Flax Seed	Honey	Ginger
Magnesium Sulphate	Flax Seed	Omega Fatty Acid	Dark Chocolate	Apple Cider Vinegar
Methionine	Fish Oil	Whey Protein	Flax seed	Cinnamon
Milk Thistle	Evening prime rose oil	BCAA	Milk and Milk Products	Pine Apple
MSM	Ferum Phos	Collagen	Calcium Salts	Oranges

	Fright			Digest	
Mahabh uta	Eleme	Secretions	Period	Salt	Secretions
Sky	Merc	Brain	Jan- Feb	Sodium	Pituitary
Air	Alu	Thyroi	Nov-Jan	Calcium	Para Th
Fire	Sil	Lung	Sug-Oct	Potassium	Thymus
Water	Phos	Adrenal	June-Aug	Ferrum	Pancreas
Earth	Sulph	Testes	Mar-May	Magnes	Ovaries

	Fight			Rest	
	Fright			Digest	
Mahabh uta	Eleme	Secretions	Period	Salt	Secretions
Sky	Merc	Brain	Jan- Feb	Sodium	Pituitary
Air	Alu	Thyroi	Nov-Jan	Calcium	Para Th
Fire	Sil	Lung	Sug-Oct	Potassium	Thymus
Water	Phos	Adrenal	June-Aug	Ferrum	Pancreas
Earth	Sulph	Testes	Mar-May	Magnes	Ovaries
	Fight			Rest